T0212110

SCIENCE AND THE BUILDING OF A NEW JAPAN

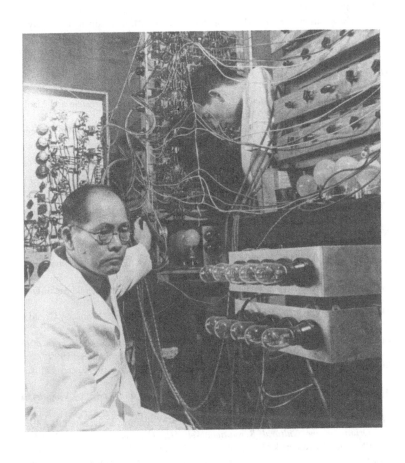

STUDIES OF THE WEATHERHEAD EAST ASIAN INSTITUTE, COLUMBIA UNIVERSITY

The Weatherhead East Asian Institute is Columbia University's center for research, publication, and teaching on the modern Asia Pacific region. The Studies of the Weatherhead East Asian Institute were inaugurated in 1962 to bring to a wider public the results of significant new research on modern and contemporary Asia Pacific.

SELECTED TITLES

(Complete list at: www.columbia.edu/cu/weai/publications.html)

Gutenberg in Shanghai: Chinese Print Capitalism, 1876–1937, by Christopher A. Reed. UBC Press, 2004

Japan's Colonization of Korea: Discourse and Power, by Alexis Dudden. University of Hawai'i Press, 2004

Divorce in Japan: Family, Gender, and the State, 1600–2000, by Harald Fuess. Stanford University Press, 2004

The Communist Takeover of Hangzhou: The Transformation of City and Cadre, 1949–1954, by James Gao. University of Hawai'i Press, 2004

Taxation without Representation in Rural China, by Thomas P. Bernstein and Xiaobo Lü. Modern China Series, Cambridge University Press, 2003

The Reluctant Dragon: Crisis Cycles in Chinese Foreign Economic Policy, by Lawrence Christopher Reardon. University of Washington Press, 2002

Cadres and Corruption: The Organizational Involution of the Chinese Communist Party, by Xiaobo Lü. Stanford University Press, 2000

Japan's Imperial Diplomacy: Consuls, Treaty Ports, and War with China, 1895–1938, by Barbara Brooks. Honolulu: University of Hawai'i Press, 2000

China's Retreat from Equality: Income Distribution and Economic Transition, Carl Riskin, Zhao Renwei, and Li Shi, eds. M.E. Sharpe, 2000

Nation, Governance, and Modernity: Canton, 1900–1927, by Michael T.W. Tsin. Stanford: Stanford University Press, 1999

Assembled in Japan: Electrical Goods and the Making of the Japanese Consumer, by Simon Partner. University of California Press, 1999

Civilization and Monsters: Spirits of Modernity in Meiji Japan, by Gerald Figal. Duke University Press, 1999

The Logic of Japanese Politics: Leaders, Institutions, and the Limits of Change, by Gerald L. Curtis. New York: Columbia University Press, 1999

Contesting Citizenship in Urban China: Peasant Migrants, the State and Logic of the Market, by Dorothy Solinger. Berkeley: University of California Press, 1999

Bicycle Citizens: The Political World of the Japanese Housewife, by Robin LeBlanc. Berkeley: University of California Press, 1999

Alignment despite Antagonism: The United States, Japan, and Korea, by Victor Cha. Stanford: Stanford University Press, 1999

China's Transition, by Andrew J. Nathan. New York: Columbia University Press, 1997

The Origins of the Cultural Revolution, Vol. III, The Coming of the Cataclysm, 1961–1966, by Roderick Macfarquhar. New York: Columbia University Press, 1997

Japan's Total Empire: Manchuria and the Culture of Wartime Imperialism, by Louise Young. Berkeley: University of California Press, 1997

Honorable Merchants: Commerce and Self-Cultivation in Late Imperial China, by Richard Lufrano. Honolulu: University of Hawai'i Press, 1997

SCIENCE AND THE BUILDING OF A NEW JAPAN

Morris Low

Johns Hopkins University

SCIENCE AND THE BUILDING OF A NEW JAPAN
© Morris Low, 2005.

Softcover reprint of the hardcover 1st edition 2005 978-1-4039-6831-9

First published in 2005 by
PALGRAVE MACMILLAN™
175 Fifth Avenue, New York, N.Y. 10010 and
Houndmills, Basingstoke, Hampshire, England RG21 6XS
Companies and representatives throughout the world.

PALGRAVE MACMILLAN is the global academic imprint of the Palgrave Macmillan division of St. Martin's Press, LLC and of Palgrave Macmillan Ltd. Macmillan® is a registered trademark in the United States, United Kingdom and other countries. Palgrave is a registered trademark in the European Union and other countries.

ISBN 978-1-349-53055-7   ISBN 978-1-4039-7692-5 (eBook)
DOI 10.1057/9781403976925

Library of Congress Cataloging-in-Publication Data

Low, Morris.
    Science and the building of a new Japan / Morris Low.
        p. cm. – (Studies of the Weatherhead East Asian Institute, Columbia University)
    Includes bibliographical references and index.
        1. Science and state—Japan—History. 2. Research, Industrial—Japan. 3. Physicists—Japan. I. Title. II. Series.

Q127.J3L69 2005
338.952'06—dc22   2005040460

A catalogue record for this book is available from the British Library.

Design by Newgen Imaging Systems (P) Ltd., Chennai, India.

First edition: August 2005

10 9 8 7 6 5 4 3 2 1

Transferred to Digital Printing in 2013

CONTENTS

Preface   vii

List of Abbreviations and Other Major Institutions   ix

List of Illustrations   xiii

1. The Making of the Japanese Physicist   1

2. Mobilizing Science in World War II:
   Yoshio Nishina   17

3. The Impact of the Allied Occupation: Nishina
   and Nakasone   45

4. Physicists on the Left: Sakata and Taketani   73

5. The Politics of Pure Science: Yukawa and
   Tomonaga   105

6. Corporate Science: Sagane   143

7. Science on the International Stage:
   Hayakawa   169

Conclusion   197

Notes   201

Glossary   253

Index   255

PREFACE

I thank my family, friends, and colleagues for their encouragement and support over the years. This book is the result of research conducted at a number of institutions including Griffith University, Nagoya University, Waseda University, the University of Sydney, Monash University, the Australian National University, and Johns Hopkins University. It was completed at the University of Queensland. I owe special thanks to Professor Roy MacLeod at the University of Sydney and Professor Tessa Morris-Suzuki at the Australian National University for their guidance, friendship, and warm support over the past two decades. During that time, the following people have been important in helping shape the pages that follow: Emeritus Professor Laurie Brown (Northwestern University), Professor James Bartholomew (Ohio State University), Professor Shigeru Nakayama (Kanagawa University), the late Professor Satio Hayakawa (Nagoya University), and Mrs. Michiko Hayakawa, the late Mrs. Nobuko Sakata, Professor Yōichi Fujimoto (Waseda University), the late Dr. Mituo Taketani, Professor Hitoshi Yoshioka (Kyushu University), Professor Robert Kargon (Johns Hopkins University), Emeritus Professor Lawrence Badash (University of California, Santa Barbara), and anonymous referees who have provided helpful comments on the manuscript. I am also grateful to Professor Carol Gluck and Ms. Madge Huntington at Columbia University who were interested in including this project in the Studies of the Weatherhead East Asian Institute from early on. I thank Dr. Hidehiko Tamaki (Nishina Memorial Foundation) and Ms. Susan Snyder (Bancroft Library, University of California, Berkeley) for permission to use correspondence held in their respective collections, and the staff of the National Diet Library for providing access to Occupation-period materials. Special thanks must go to Dr. Spencer Weart (director, Center for History of Physics, American Institute of Physics) for encouraging me, several years ago, to use the rich resources of the Emilio Segrè Visual Archives, and Ms. Heather Lindsay, assistant librarian/photo administrator, for making it all happen.

I also wish to express my warm thanks to Anthony Wahl and Heather Van Dusen at Palgrave Macmillan, Maran Elancheran at Newgen Imaging Systems, and the copy editor who has worked on this project, my second book with them. Without their interest, support, and patience, this book would not have been possible.

As with most books dealing with Japan, a few words are required on romanization and other conventions followed in this book. Although family names precede given names in Japan, many Japanese (especially scientists) write their names in reverse order when writing in English. To conform with this, I have used the Western order throughout. The romanization of Japanese follows that found in the Kenkyūsha Japanese–English dictionary. I have avoided use of hyphens unless absolutely necessary. The names of some people are romanized differently, because they are widely known by a certain spelling. Macrons over elongated vowels in well-known placenames, such as Tokyo, Kyoto, and Osaka, have been dropped. Verbs are spelt with "ize" instead of with "ise." In quotations, the original spelling has been retained.

Morris Low

Abbreviations and Other Major Institutions

Abbreviations

ABCC	Atomic Bomb Casualty Commission
AEA	(UK) Atomic Energy Authority
AEC	(U.S.) Atomic Energy Commission
AIP	American Institute of Physics
BWR	boiling water reactor
CCD	Civil Censorship Detachment, SCAP
CEA	Commissariat à l'Energie Atomique
CIE	Civil Information and Education Section, SCAP
CIS	Civil Intelligence Section, SCAP
CNRS	Centre National de la Recherche Scientifique, France
COSPAR	Committee for Space Research (1959–)
CRC	Cosmic Ray Researchers Council
ESS	Economic and Scientific Section, SCAP
FEC	Far Eastern Commission, Washington
FEO	(Japan) Federation of Economic Organizations (Keidanren)
GHQ	General Headquarters, SCAP
IAEA	International Atomic Energy Agency
ICSU	International Council of Scientific Unions
IGY	International Geophysical Year [1957–1958]
IIS	Institute of Industrial Science, University of Tokyo
INS	Institute for Nuclear Study, Tokyo (1955–1997) In 1997, merged with National Laboratory for High Energy Physics (KEK) to become High Energy Accelerator Research Organization

ISAS	Institute of Space and Astronautical Science [1981–] Formerly Institute of Space and Aeronautical Science, University of Tokyo (1964–1981).
JAEC	Japan Atomic Energy Commission (est. 1956)
JAERI	Japan Atomic Energy Research Institute (1956–)
JAFC	Japan Atomic Fuel Corporation (1957–)
JAIF	Japan Atomic Industrial Forum (1956–)
JAPCO	Japan Atomic Power Company
JASL	Japan Association for Science Liaison
JNSDA	Japan Nuclear Ship Development Agency (1963–)
JPDR	Japan Power Demonstration Reactor
JSC	Japan Science Council (1949–)
KEK	(Kō Enerugi Butsurigaku Kenkyūjo) National Laboratory for High Energy Physics, Tsukuba (1971–1997) In 1997, merged with INS to become High Energy Accelerator Research Organization
LDP	Liberal Democratic Party (1955–)
LWR	light water reactor
Minka	(Minshushugi Kagakusha Kyōkai) Democratic Scientists' Association (est. 1946)
MIT	Massachusetts Institute of Technology
MITI	Ministry of International Trade and Industry In 2001, renamed Ministry of Economy, Trade and Industry
NASA	(U.S.) National Aeronautics and Space Administration
NASDA	National Space Development Agency of Japan (1969–)
NATO	North Atlantic Treaty Organization
NRC	National Research Council, Japan
NRS	Natural Resources Section
NTT	Nippon Telegraph and Telephone Public Corporation
OSRD	(U.S.) Office of Scientific Research and Development
PH&W	Public Health and Welfare
PNC	Power Reactor and Nuclear Fuel Development Corporation (est. 1967)
PWR	pressurized water reactor
Riken	(Rikagaku Kenkyūjo) Institute of Physical and Chemical Research (1917–, renamed "Science Research Institute" in 1951, but now known by former name.)
R&D	research and development
REKS	Research Group of Elementary Particles in Kansai

REKT	Research Group of Elementary Particles in Kantō
RIFP	Research Institute for Fundamental Physics Kyoto University (1953–)
ROKK	(Roketto Kansoku Kyōgikai) Rocket Observation Council
SCAP	Headquarters of the Supreme Commander of the Allied Powers
SCNR	Special Committee for Nuclear Research, JSC (1952–?)
SJC	(Soken Junbi Chōsa Iinkai) KEK Preparatory Committee
STA	Science and Technology Agency (1956–2001) In 2001, merged with Ministry of Education, Science, Sports and Culture to become Ministry of Education, Culture, Sports, Science and Technology
STAC	Scientific and Technical Administration Committee
UNESCO	United Nations Educational, Scientific, and Cultural Organization
USIS	U.S. Information Service
WAWF	World Association of World Federalists
ZETA	Zero Energy Thermonuclear Assembly

OTHER MAJOR INSTITUTIONS

Institute of Plasma Physics, Nagoya University (1961–1989). Renamed the National Institute for Fusion Science and later moved to Toki, Gifu

Japan Society for the Promotion of Science (Nihon Gakujutsu Shinkōkai) (1932–1945, then reorganized)

Naval Technical Research Institute (1923–1945)

Science and Technology Council (Kagaku Gijutsu Kaigi) (1959–)

Association of Experimental Nuclear Physicists (Genshikaku Danwakai)

ILLUSTRATIONS

1.1 Growth in first-year undergraduate science enrolments
according to specialization (1958–1970). Adapted from
Tetu Hirosige, *Kagaku no shakaishi: Kindai Nihon no
kagaku taisei* (*The Social History of Science:
The Organization of Science in Modern Japan*)
(Tokyo: Chūō Kōronsha, 1973), p. 294   12

3.1 Yoshio Nishina and I.I. Rabi at a welcome party for the
U.S. Scientific Advisory Group, Tokyo, December 1948.
Source: Nishina Memorial Foundation. Courtesy,
AIP Emilio Segrè Visual Archives   53

4.1 Shōichi Sakata and Mituo Taketani at the Japan Academy,
Ueno, Tokyo, where a General Meeting of the Science
Council of Japan was held on April 28, 1951.
Photograph: Ken Domon. Courtesy, late Mrs. Nobuko
Sakata and Prof. Fumihiko Sakata   91

5.1 Robert Oppenheimer, Hideki Yukawa, and
Sin-itirō Tomonaga, November 1949, at the
Institute of Advanced Studies, Princeton, just after
the announcement of Yukawa's Nobel Prize. Courtesy,
AIP Emilio Segrè Visual Archives, Yukawa Collection   111

5.2 Hideki Yukawa surrounded by reporters on the
occasion of his return to Japan after receiving
the Nobel Prize, 1950. *Source*: Mainichi
Shinbun-sha. Courtesy, AIP Emilio Segrè
Visual Archives, Yukawa Collection   112

5.3 Sin-itirō Tomonaga, center, with hand raised,
seated beside Nobuyuki Fukuda at a seminar at the
Okubo branch of the Physics Department, Tokyo
Bunrika University, Tokyo, 1948. Photograph:
Shunkichi Kikuchi. *Source*: University of Tsukuba,

Tomonaga Memorial Room. Courtesy, AIP Emilio
Segrè Visual Archives 123
5.4 Major elements of government R&D policy-making
structure, ca. 1950s–1960s. Adapted from Ministry of
Education, Science and Culture, *An Outline of the
University-Based Research System in Japan*
(Tokyo, 1983), p. 3 128
5.5 Hideki Yukawa seated in his home, doing calligraphy,
March 1962. Courtesy, AIP Emilio Segrè Visual
Archives, Yukawa Collection 134
5.6 Sin-itirō Tomonaga wearing a kimono and
standing outdoors in his garden, ca. August 1970.
Photograph: Hiroshi Chiba. Courtesy, AIP Emilio
Segrè Visual Archives 136
6.1 Hideki Yukawa, Ryōkichi Sagane, and John Cockcroft in
Hakone National Park, Japan, November 22, 1958.
Courtesy, AIP Emilio Segrè Visual Archives,
Yukawa Collection 161
7.1 Back row (standing): Jun J. Sakurai, Chūshirō Hayashi,
Mituo Taketani, and Satio Hayakawa. Seated:
Seitarō Nakamura (second from the left) and
Hideki Yukawa (center right). February 1955 in
Yukawa Hall, Kyoto University. Courtesy,
AIP Emilio Segrè Visual Archives, Yukawa Collection 177
7.2 Sumi Yukawa, Satio Hayakawa, Richard P. Feynman,
Hideki Yukawa, Kōichi Mano, and Minoru Kobayasi,
during a visit to Yukawa Hall, Kyoto University,
summer of 1955. (Image has been reversed.) Courtesy,
AIP Emilio Segrè Visual Archives, *Physics
Today* Collection 178
7.3 Organization of fusion research, ca. 1950s–1960s.
Adapted from Akira Oikawa, "History and
Organization of Fusion Research and Development
in Japan," *Fusion Nuclear Technology*, vol. 17, no. 2
(March 1990), pp. 232–35, esp. p. 233 185
7.4 Major elements of the organization of government
space R&D, ca. 1950s–1960s 192

CHAPTER 1

THE MAKING OF THE
JAPANESE PHYSICIST

The image of a hi-tech samurai has often been invoked to describe Japan's post–World War II economic success. But such references to the role of Japan's warrior class go back to the beginning of the twentieth century. "Scratch a Japanese of the most advanced ideas, and he will show a samurai"[1]—so wrote Inazō Nitobe in his classic text, *Bushido: The Soul of Japan*, first published in 1900 and then in a revised form in 1905, the year of Japan's victory in the Russo-Japanese War (1904–1905). Nitobe, who studied politics and international relations at Johns Hopkins University in Baltimore during the years 1884–1887, develops an argument, along the lines that "What Japan was she owed to the samurai."[2] He suggests that the samurai became an ideal for the Japanese and that the spirit of bushidō permeated all social classes.[3]

"Bushidō," which translates as "The Way of the Warrior," refers to an ethos that emerged in the seventeenth century. It emphasized moral uprightness, courage, benevolence, sincerity, honor, loyalty, and self-discipline. Confucian scholars promoted these qualities during the Tokugawa period (ca. 1603–1868) as part of an ethical code for the samurai. This was in turn used by the samurai to justify their privileged position in society as part of an elite, hereditary class. Samurai values were used as a means of national integration during the Meiji period, despite the fact that only a small minority of the Japanese population actually were samurai. This didn't stop the country from using such ideals as a gender construct. In the late 1950s, they were called on again by conservative males, such as the politician Yasuhiro Nakasone, who longed for a return to traditional values.[4]

The promotion of the image of businessmen as modern-day warriors has drawn attention away from the role of scientists, especially physicists, in the making of a new Japan in the postwar period. While scientists in many fields sought to contribute to reconstruction, the strategic importance of nuclear weapons lent extra leverage to physicists as actors in the political arena. Although Japan did not seek nuclear arms capability after the war, it did establish a civilian nuclear power program for which physicists were to prove useful. In the minds of young science students in postwar Japan, the physicists discussed in this book occupied a central place.

What sort of people were they? The development of science in the West was supported by group values, which many associated with gentlemen. In nineteenth-century England, "being a gentleman was an important and typical characteristic of the man of science."[5] We shall see in this book certain parallels between the interests and behaviors associated with the samurai class and that of men of science.

Adoption was common amongst samurai families. It helped younger sons find homes and achieve financial independence; placed them in positions of influence; and provided upward social mobility not only for the sons but sometimes for the adopting family as well. It is estimated that in the seventeenth century, up to a quarter of all samurai families may have had adopted sons. This practice became even more frequent in the eighteenth and nineteenth centuries, with more than a third of families likely to have turned to adoption of younger sons of samurai families as a means of continuing the family line and reproducing their social class.[6]

The physicists discussed in this book tended to come from elite family backgrounds where practices such as adoption were not frowned upon. We shall see that for two of them, the physicists Hideki Yukawa (born Hideki Ogawa) and Ryōkichi Sagane (born Ryōkichi Nagaoka), adoption ensured that they gained some financial security and, at the same time, helped perpetuate the family lines of some well-to-do families. Thus, in some respects, the making of the Japanese physicist also involved the making of a social class. Of course, there were exceptions, such as Mituo Taketani, and there was generational change. The generation of physicists born between 1935 and 1955 sometimes found it difficult to bridge the class differences between themselves and their higher social class teachers.[7]

Qualities associated with the samurai served as a cultural resource, a construct, which helped shape the postwar public persona of physicists such as the Nobel Prize winners Hideki Yukawa and Sin-itirō Tomonaga. The tendency to view samurai ideals as something best forgotten and relegated to Japan's feudal past, hides the reality that such ideals shaped the persona of Japan's leading scientists. The privileged background of physicists such as Shōichi Sakata seems to have influenced their worldview, instilling in them a desire to help those less fortunate and to create a better world. Socioeconomic background did not prevent less wealthy physicists from using this as an ideal. For left wing intellectuals such as Mituo Taketani, science could help build a better world by contributing to an anti-fascist "world culture." Such ideas can be seen in less radical form in the internationalism and public mindedness of later physicists such as Satio Hayakawa. These qualities seem to resonate with characteristics previously associated with the samurai.

Historians of Japanese science have, in the past, explored such ideas, especially in terms of the samurai background of early scientists. But the idea of the samurai as an identity construct that somehow converged with the ideals and norms of being a Western scientist has not been adequately explored. We can point to similarities in the behavior of German and Japanese scientists; yet, at the same time, the Japanese have arguably displayed values different from those of Western scientists.[8] The introduction of Western science, the establishment of scientific communities, and their relationship with society show value orientations quite different to those of the West. How do we reconcile these seemingly opposing views? It is possible to argue that at the end of the nineteenth century, the Japanese had intellectual traditions that gave scholars, "a range of cognitive perceptions and cultural predispositions" relatively compatible with scientific progress.[9] Educated Japanese people took up science in the nineteenth century in a manner reminiscent of the samurai pursuit of learning. This bonded science with the public sphere and brought about the notion of science in service to the state.

We need only to return to Nitobe to see how this was possible. He, like the physicists discussed in this book, had an international reputation. Born the son of a wealthy samurai (from Morioka), he studied overseas—at Johns Hopkins University in the United States and at Halle University in Germany—held an academic position at Tokyo University, and was involved in international organizations (undersecretary general of the League of Nations and founder of the forerunner of UNESCO).[10] Although he was more interested in economics and agriculture rather than science per se, he provides an example of an international Japanese scholar, involved in state building, to whom physicists can be usefully compared, in that there are some shared aspects of their lives. Samurai ideals and the notion of public service and social responsibility seem to have shaped the lives of the physicists discussed in this book.

I THE PREMODERN VISION OF SCIENCE

During the Tokugawa period, the samurai class (5–6% of the population) devoted their energies to civil administration and the literary and martial arts. Learned samurai were the intellectuals during this time, highly cultured and knowledgeable in the Confucian classics.[11] The scholarly concerns of these "public men" were of a nonutilitarian nature. The samurai bureaucracy saw Chinese mathematics and astronomy as routine skills and relegated them to the appropriate occupational category in the social hierarchy.[12] "Learning" tended to be equated with the intellectual activities of the samurai, even when the scholar was not of the warrior class.

In the seventeenth century, commoners were most active in the physical sciences, such as astronomy and calendrical studies. From the eighteenth century, samurai increasingly turned to the sciences and by the mid-eighteenth century, had become the dominant group. This was somewhat similar to seventeenth-century England where the scientist was a cultured gentleman, not an artisan. In Thomas Sprat's *History of the Royal Society* (1667), it is argued that

> there is also some privilege to be allow'd to the generosity of their spirits, which have not been subdu'd, and clogg'd by any constant toyl, as the others. Invention is an Heroic thing, and plac'd above the reach of a low, and vulgar Genius.[13]

The situation in the field of mathematics in Japan, however, was different. Samurai showed interest in mathematics early in the Tokugawa period, but the mercantile and agrarian classes increasingly took control of the field. The education of commoners tended to be pragmatic, and with the growth of the economy during the eighteenth century, skills in arithmetic became important. Merchants increasingly co-opted mathematics for commerce, and samurai grew to despise what had become associated with the people to whom they had become financially indebted.[14]

Despite the important role of merchants in the development of technical skills, wealthy merchants and other commoners aspired to a formal education in the samurai mode. Paradoxically, this education was not particularly suited to natural knowledge as it emphasized moral values, an understanding of government and responsibilities to society.[15] But education along samurai lines enabled social advancement, and the "usefulness" of what was learned was a secondary consideration. In spite of such ideals, the education of commoners tended to be more pragmatic, not least because the growth of the economy demanded skills in mathematics. In the late Tokugawa period, while there were public institutions for science that served the needs of the state, private individuals (physicians and founders of academies) dominated the overall scientific movement from the eighteenth century onward. One would do well to remember, however, that Tokugawa scholars were "confined to a cage called government," spending "their time in anguish within the small universe which this cage created."[16] The state helped shape their worldview.

II   THE MEIJI VISION OF SCIENCE

During the Tokugawa period, talented people were attracted to the pursuit of natural knowledge. While their intellectual contribution may not have been particularly significant in terms of Western science, clear continuities exist between the backgrounds and class of people who pursued science during the Tokugawa period and those who took up science in the Meiji period (1868–1912).[17] The emergence of the modern state in Meiji Japan was accompanied by the emergence of a "public" sphere, which produced intellectuals we can refer to as public men. This public sphere went beyond the official and private activities of the Japanese. While "publicness" related to public activity, "insiders" (with links to large organizations) were esteemed more highly than "outsiders" (dissidents, independents) with no such links. Public thus tended to be viewed as being associated with the state, and the intellectual elite of both persuasions derived their status as experts in their chosen areas of imported knowledge.[18]

After the Meiji Restoration of 1867–1868, the samurai military elite lost their traditional occupations in the feudal bureaucracy and had to turn elsewhere for their income. The study of medicine and agriculture was already the domain of the offspring of physicians and wealthy farmers. Law was not considered an activity fit for samurai, and so many turned to science and engineering as promising professions. These samurai scholars were well suited to serving the state, coming as they did from a tradition of civil service. This elite group was more of government officials than independent-minded intellectuals. Rather than attempting to think of new ideas, they were busy conforming, as quickly as possible, to conventional Western views of nature.[19] One could argue that the traditional conception of nature ill-prepared them for creative thinking.

In the mid-nineteenth century, scholars tried to reconcile Neo-Confucian concepts of man and nature by equating Western science with "kyūri," one of the concepts of Chu Hsi philosophy. Kyūri is the process of perceiving and comprehending the "ri" ("li" in Chinese), which permeates all nature. With the widespread adoption of Western science in the Meiji period, the term "kyūrigaku" for physics became "butsurigaku." The component ri is still found in many terms to denote science, such as "rigakubu" ("School of Science").[20]

Germany was promoted as an academic model during this time—one of the many organizational models that Japan introduced during the Meiji period.[21] Interest in German scholarship intensified in the 1880s and Japanese scientists studying in Europe gravitated toward Germany. Government ministers looked up to German higher education as a model. It would only be after World War II that the American style of graduate school education have an impact.[22] After Japan's victory in the Sino-Japanese War (1894–1895), access to European industrial research facilities became more difficult, compelling Japanese firms to establish research facilities of their own.[23]

Education and research in the physical sciences in the Meiji (1868–1912) and Taishō (1912–1926) periods were reliant on the public sector and scientists were, as a result, considered government officials. The majority of physicists tended to have a samurai background, and almost all went on to become professors. Early physicists tended to come from the community of Confucian-trained officials, whereas chemists were offspring of physicians trained in Chinese-style medicine. This reflects how the official class had greater access to the physical sciences, and it has been found that Tokyo tended to act as a locus for such know-how. The Japanese physicist is thus epitomized as a member of the social elite, schooled in the Confucian classics, highly urban, likely to take up a position as a university professor, and reliant on the government for support of his activities.[24]

The education of Japan's early physicists was based on the education of a samurai. A samurai would traditionally study the five major Chinese Confucian classics. After the Meiji Restoration, this emphasis on Chinese studies declined, giving way to Western learning.[25] In the 1870s and 1880s, great emphasis was placed on science and technology to facilitate Japan's industrialization and to channel bright students away from potentially subversive subjects. The "Proposals on Education," which were submitted to the Emperor in 1879, directed that: "Upper level students should be schooled in the sciences; they should not be drawn off into discussions of politics."[26]

In the Meiji period, science and technology were mobilized under an ideology aimed at building a nation-state. The potential collision between Confucianist morals and Western science was cleverly avoided by the propagation of slogans such as "Eastern morality, Western art forms."[27] The main aim was to catch up with Western science as quickly as possible, so that such knowledge could be used in the building of a modern state. Priority was given to "public sciences," such as geographical surveys, standardization of units of measurement, meteorology, sanitation, printing, communications, and transportation. Pilot plants for the textile industry were established and new industries were nurtured.[28]

Foreign teachers in the physical sciences tended to be either British or American.[29] The role of British ideologies can be seen in the activities of the Imperial College of Engineering (known in Japanese as "Kōgakuryō" and later "Kōbu Daigakkō"), established in 1873 under the control of the Ministry of Public Works and under the directorship of Henry Dyer. Dyer established engineering programs that would later be emulated back in Britain.[30] As in physics, early engineering students tended to be young samurai who would go on to become technocrats and, later, politicians.[31]

It was only in 1875 that the Tōkyō Kaisei Gakkō, a forerunner of Tokyo University, offered the first, independent advanced course in physics. Two years later, in 1877, a university system of scientific and technical education was established with the creation of the Kōbu Daigakkō (School of Engineering) and the University of Tokyo (centered upon a school of science that had originally been the Tōkyō Kaisei Gakkō). It was also in 1877 that the Tokyo Mathematical Society was established, changing its name to the Tokyo Mathematico-Physical Society (Tōkyō Sūgaku Butsuri Gakkai) in 1884. In 1879, Kenjirō Yamagawa became the first Japanese to be appointed to a chair in physics at the University of Tokyo, and by 1882, the first three students with a major in physics had graduated from that institution. Such rapid institution building is indicative of the rapidity of change in Japan during the 1870s.

The Ministry of Public Works was abolished in 1886 and control of the School of Engineering was transferred to the Ministry of Education, which incorporated it into the newly organized Tokyo Imperial University, established in 1886. The University thus came to be characterized as having a large School of Engineering.[32] Kyoto Imperial University soon followed in 1897, and Tohoku Imperial University in 1907. Thus, by the turn of the century, there were three universities producing physicists.[33] The Tokyo Imperial University produced 104 physics graduates by the year 1900 and around 4 or 5 per year after that. Kyoto Imperial University produced its first graduates majoring in physics in the year 1902.[34]

Despite the impressive progress made in institutional structure, the German Erwin Baelz is known to have complained that in Meiji Japan "the origin and nature of science is largely misunderstood. Science is seen as a machine. . . ."[35] This instrumental approach to science was exacerbated by the independent transfer of various specialized fields of knowledge without due consideration of their inter-relatedness. In the rush to learn and adopt the latest developments in science, little concern was shown for preserving Japanese traditions. It has often been claimed by the Japanese that there was little conflict between the two strands of knowledge as neither engaged the other. One, it was said, merely superimposed itself on the other, acting like a new veneer.[36] Furthermore, the rapid introduction of Western science resulted in neglect of the original philosophical and cultural context of science. It can, however, be argued that Japanese culture (including the ideal of the samurai) provided a resource from which a Japanese science, different in its emphasis on the practical and utilitarian, and different in its philosophical roots, could be shaped. Science became tied to the pursuit of national interests and profit.

For example, for the Meiji physicist Kenjirō Yamagawa, knowledge and education served the state. Science as an intellectual pursuit was not for him: "There are many Japanese who have died out of loyalty, but I know of no one who has died for the truth."[37] Instead, scholars showed respect for their elders and had a tendency toward factionalism. Such characteristics did not encourage strong individuality or a free exchange of information or opinions. Knowledge was still considered to be private property and not part of the public domain.[38]

The emergence of the concept of public science (science for public service) during the Meiji program of modernization was arguably a precondition for Japan's industrialization. Despite the lack of a conceptual divide between science and technology, institutionalization meant compartmentalization into various disciplines along traditional Western lines with "scientists" and "engineers." The public science and technology practiced by the samurai elite was Western, science-based, and located in the university. Science and technology in the private sector tended to be more traditional and not based along Western science principles. This was evident in sake brewing, ceramics, carpentry, and fishing.[39]

III BUILDING A WESTERN
VISION OF SCIENCE

The institutionalization of science in the following decades placed the pursuit of knowledge in a Western context of professional societies and journals, but never too far away from the demands of the state. Aspects of this new Western scientific discourse were not unlike the samurai's approach to learning. Participation in it was prestigious, enabled social advancement, had connotations of public service, and involved, at times, a disdain for the utilitarian. As in the Tokugawa period, such an approach ran counter to the traditional stance of the private sector, which continued to view knowledge as private property—an engine of economic growth enabling the Japanese to catch up with the West. No matter how removed from commercial realities, universities, nevertheless, served an important role in providing science and engineering graduates for industry. Public science in universities may have conformed to Western science norms, but it represented only the interface between science, society, and the international community. Much scientific activity occurred in the private sector, away from the gaze of the public and international assessment. In the two decades before Japan's entry into World War II, many research laboratories were established by private enterprises, especially in shipbuilding, electrical machinery, and equipment. In 1930, there were over 300 corporate laboratories, around 20 government laboratories, and 70 attached to imperial universities or run by local government. The total sum of such research organizations increased to about 700 by 1935.[40]

During the period from 1898 to 1956, the number of doctorates of science ("hakasegō") awarded at Japanese universities grew exponentially.[41] This trend is also reflected in membership of the Physico-Mathematical Society of Japan and in the growth of physics. (Note that the Physical Society of Japan was only established in 1947.) The first generation of physicists provided a base of teaching personnel for the domestic production of Japanese physicists. This generation of some 15 members consisted of foreign teachers and Japanese physicists who had been trained abroad and who returned to Japan during the Meiji period. This period of largely "colonial science" was succeeded by a second generation of their students who numbered around 20. This second wave emerged in the late 1870s, and continued until the beginning of the Shōwa period in the mid-1920s. This generation of Japanese practitioners taught a third generation of Japanese students in the Japanese language, paving the way for a more national science and establishing a community of physicists that has doubled every 7 years since 1900.[42] Such graduates would go on to become members of the Japanese scientific community.

During and after World War I, a number of research facilities were established including the Institute of Physical and Chemical Research (abbreviated in Japanese as "Riken") in Tokyo, the largest pre–World War II scientific research organization in Japan. It was analogous to the Imperial Institute of Physics and Technology in Germany (established in 1887).[43] The charter of Riken reflects its purpose:

> To promote the development of industry, the Institute of Physical and Chemical Research shall perform pure research in the fields of physics and chemistry as well as research relating to various possible applications. No industry, whether engaged in manufacturing or agriculture, can develop properly if it lacks a sound basis in physics and chemistry. In our country especially, given its dense population and paucity of industrial raw materials, science is really the only means by which industrial development and national power can be made to grow. The Institute has thus set forth this important mission as its basic objective.[44]

Riken was formed in 1917 with funds from the government and the private sector. As its name suggests, the institute was, at least until 1921, divided into two divisions: physics and chemistry. To overcome rivalry between the two, they were made independent of each other with regard to funding, staffing, and research. Considerable administrative and intellectual freedom facilitated both theoretical and applied research. The head of physics was Hantarō Nagaoka. From 1922, a laboratory system was established, which enhanced the institute's research. The man who was responsible for this structure was Masatoshi Ōkōchi, director of Riken from 1921 to 1946. (Viscount) Ōkōchi was a descendant of a feudal lord and well connected. He abolished the divisions of physics and chemistry, setting up the above-mentioned system of chief-researchers who would be in charge of their own laboratories. There were initially 14 chief-researchers, but by the early 1940s, this had increased to 33 with a total of around 1800 researchers. This was in contrast to a total of 200 members in 1918. Riken attracted many promising young researchers, many of whom used the facilities there for research toward a doctoral degree.[45]

The interruption of the academic exchange between Europe and Japan during World War I helped to create a more "Japanese" science with links to the Japanese industry. By the end of the Taishō period, during the 1920s, the beginnings of a Japanese physics had been born. The 1930s would prove to be a particularly formative period.

Despite the wide interest in science, there were only three schools of science in Japanese universities in 1926, the first year of the Shōwa period. The school at Tokyo dated back to 1877, Kyoto to 1897, and Tohoku to 1911. (Kyushu Imperial University was established in 1911 with programs only in medicine and engineering.) Before long, however, science schools were created at Hokkaido (1930), Osaka (1933), Kyushu (1939), and Nagoya (1942) imperial universities. Tokyo Bunrika University and Hiroshima Bunrika University also had a major science focus.[46]

Despite the vigor of research activities at Riken, the formal teaching of physics largely rested with the seven imperial universities: Tokyo, Osaka, Nagoya, Kyoto, Hokkaido, Kyushu, and Tohoku. Some of the physics graduates from these institutions went on to become full-fledged researchers at institutions such as Riken. By 1940, university science graduates were over double the number in the first year of the Shōwa period, 1926. Five years later, it was over double that again.[47]

Soon after the end of World War II, the Scientific and Technical Division of the Economic and Scientific Section (ESS) of the Supreme Commander for the Allied Powers (SCAP) did a stock-take of scientific and technical personnel in Japan.[48] Between 1931 and 1940, around 1159 doctorates in science, technology, agriculture, and forestry were awarded in Japan. As for the United States during the same period, there were 11,443 PhD degrees. This means that the rate of production of highly trained scientists and engineers in Japan, per head of population, was about one-fifth that in the United States.[49] Not bad for a latecomer.

The surprise launch of the first artificial satellite, Sputnik I, by the Soviet Union on October 4, 1957 marked the beginning of the space race with the United States. Other nations, too, were spurred into taking action, with Japan dramatically increasing its levels of scientific manpower. The Ministry of Education announced a plan that same year to increase the number of students admitted into science and engineering programs, aiming to increase the entrance quota from 22,000 to 30,000 by 1960. Having achieved this, the Ministry further launched a plan in 1961 to increase the number of new student places by an extra 20,000 by 1964; this was in order to meet the strong demand for scientific manpower from the industry.[50]

It is against this background of a massive increase in the number of students majoring in science that we see amazing growth in the

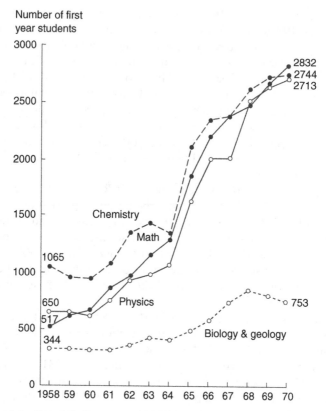

Figure 1.1   Growth in first-year undergraduate science enrolments according to specialization (1958–1970).

Adapted from Tetu Hirosige, *Kagaku no shakaishi: Kindai Nihon no kagaku taisei* (*The Social History of Science: The Organization of Science in Modern Japan*) (Tokyo: Chūō Kōronsha, 1973), p. 294.

number of students studying mathematics and physics. Numbers increased 5.5 and 4.3 times, respectively, over the period 1958–1970 (see figure 1.1). This was in contrast to chemistry, which grew less strongly, and biology and geology, which when combined grew only 2.2 times over the same period. Mathematics and physics majors found ready jobs in electronics, space research, nuclear power, and information technology.[51] In the United States, the production of young physicists also provided a reserve labor force of potential weapons makers, and was more closely tied to the Cold War.[52]

IV   Case Studies

In the chapters to follow, we see how like many physicists of his day, Yoshio Nishina (1890–1951) came from a privileged background.

Shōichi Sakata (1911–1970), too, was born into a family with a tradition of public service. While working at Nishina's laboratory, he befriended Mituo Taketani (1911–2000) who, although hardly wealthy, sought, like Sakata, to be a social spokesman for science throughout his long career. Family, education, and marriage shaped the identity of Hideki Yukawa (1907–1981), Japan's first Nobel Prize winner in physics. Yoshio Nishina proved to be a father figure for Sin-itirō Tomonaga (1906–1979), and upon his death, Tomonaga would take on many of the leadership roles that Nishina had served in. The stories of the lives and careers of these and other physicists, and how they contributed to building a new Japan after the war, is the subject of this book.

We will see that physicists were able to take on hybrid or multiple identities. As public men, they contributed to Japan's reconstruction in four critical ways—as social spokesmen, as experts, as technocrats, and as international scientists. Each of the next six chapters examines these roles. The chapters raise questions about the place of scientific expertise in Japanese society and consider separate, but, nevertheless, overlapping, domains—the scientific community and the military, the scientific community and the wider community, government science policy, the relationship between universities and the corporate sector, and the use of science for international prestige and foreign policy.

While women will appear from time to time in this book, they are on the margins. This reflects the continuing reality of the lack of women in science and engineering in Japan. This can be attributed to a number of factors including socioeconomic conditions, gender ideology, and cultural traditions of feudal Japan. Gender expectations at work in Japanese society were reinforced by those in science themselves to effectively marginalize women and exclude them from Japanese science.

Chapter 2 considers the activities of the physicist Yoshio Nishina before and during World War II, his role at the Institute of Physical and Chemical Research in Tokyo, and the Japanese atomic bomb project of which he was a leader. Nishina co-opted state discourses to achieve his own research agenda, portraying his work as part of a larger mobilization of science, which was designed to centralize science and bolster Japan's war effort. The chapter ends with the dropping of the atomic bomb on Hiroshima, and the young bureaucrat and budding politician Yasuhiro Nakasone resolving to harness science for the nation.

Chapter 3 examines the impact of the American-led democracy on the Japanese, seen again through the lives of Nishina and Nakasone. Nishina attempted to reorganize the Institute of Physical and

Chemical Research, this time engaging in the rhetoric of democracy, in order to ensure its survival under the Allied Occupation. Reflecting the fervor for greater participation in policy-making, Nishina was instrumental in establishing, in cooperation with the Occupation Forces, the Science Council of Japan. Meanwhile, Nakasone transformed himself into a successful, conservative politician.

In chapter 4, we consider the social contribution that physicists made by looking at the Marxist physicists Shōichi Sakata and Mituo Taketani. Sakata, in particular, was influenced by J.D. Bernal, and by the idea that the central planning of science could be made compatible with democracy and the needs of the public. Taketani has been one of the most prolific scientist–social critics, his writings receiving extensive exposure in the media and through the publishing house Keisō Shobō. These physicists appealed to the "public good" to promote their notion of good science, which, more often than not, implied basic research conducted in Japan.

Chapters 5 and 6 reveal how interaction with the Liberal Democratic Party, bureaucracy, and big business has been very much a part of the lives of physicists as policy-makers. The activities òf Nobel Prize physicists Hideki Yukawa and Sin-itirō Tomonaga— eminent figures in the Japan Atomic Energy Commission and the Science Council of Japan—are outlined in chapter 5. Both felt that science could contribute to society by adding to the bank of basic knowledge. They lent credibility to government bodies, scientist organizations, and peace groups. In chapter 6, we focus on Ryōkichi Sagane (1905–1969), physicist and son of the elder statesman of Japanese physics, Hantarō Nagaoka. Sagane, who studied with Ernest Lawrence at Berkeley, was instrumental in transferring cyclotron technology to Japan. After the war, Sagane cooperated with the government and big business in atomic energy. His technocratic activities can be contrasted with those of physicists such as Satio Hayakawa (1923–1992), who promoted science with the aim of establishing a name for Japan in international scholarship.

"Catching up with the West" was a useful strategy employed by physicists such as Hayakawa to argue for greater funding of university research. Chapter 7 describes the many areas of policy-making and research in which Hayakawa participated. Although his early career saw him in cosmic-ray and elementary-particle physics, his interests extended to astrophysics and space sciences, as well as nuclear fusion and plasma physics. He developed common-use research institutes, including the Research Institute for Fundamental Physics at Kyoto University, the Institute for Nuclear Study, the Institute of Space and

Astronautical Sciences (both formerly attached to Tokyo University), the Institute for Plasma Physics (formerly at Nagoya University), and the KEK High Energy Physics Laboratory at Tsukuba. While mindful of possible spinoffs, his overriding aim was to promote Japanese physics to international levels of prestige.

CONCLUSION

This chapter has outlined the making of the Japanese physicist. Many early Japanese physicists did come from samurai background and for others it served as an ideal, if not a reality. In postwar Japan, scientists employed different public discourses to link science with the state and to achieve their own goals. There was an exciting "moment" that allowed socialism, liberalism, and democracy to reenter public discourse.[53] Japanese scientists were among those who took up the new discourses, just as some had adopted the wartime state discourse to suit their own purposes. The renewed discourse of democracy enabled the Japanese public men of science to promote the role of science in the reconstruction of Japan and for the public good. Physicists tried, where possible, to place themselves in positions of influence to determine the direction of science policy. While physicists co-opted state discourses relating to Japan's war effort and postwar democracy to further their own aims, what has tended to be ignored is how their personas as public men were shaped by samurai ideals. The ideal of the samurai arguably informed their activities, despite their claims of abolishing the "feudalism" of the academic system.

The samurai originally pursued this foreign learning as an intellectual diversion rather than for utilitarian purposes, positioning it rather awkwardly within the context of Confucian values. Elite scientists in universities have continued to operate in Japanese society, somewhat removed from the mainstream and having to meet the needs of the state, away from the everyday pressure of development at all costs. Institutions such as a strong bureaucracy, the former financial conglomerates known as the "zaibatsu," and banking system were created to achieve high economic growth.[54] Science has come from a different socioeconomic context. This has resulted in a tension in the identity of Japanese physicists as public men. To what extent do they serve the public good and to what extent should they act in the interests of the state? As we shall see in the chapters to follow, it was a matter for each physicist to decide.

CHAPTER 2

MOBILIZING SCIENCE IN
WORLD WAR II: YOSHIO NISHINA

We saw, in chapter 1, how the Japanese physicist was shaped by both social and cultural factors. Indeed, the samurai "spirit" can be considered a cultural resource, a construct, which Japanese used in both peacetime and in war. World War II provides a useful window to how physicists negotiated their multiple identities and sometimes conflicting loyalties. In this chapter, we focus on Yoshio Nishina, who effectively mentored a whole generation of physicists. He is also considered the father of the Japanese atomic bomb. Although Nishina's lab lacked equipment and materials to pursue the study of the artificial disintegration of the elements, he wrote

> we cannot help to be tempted to do this sort of experiments, because there are so many problems which are awaiting their solution. It is quite possible that we come too late, but that does not matter, it is the way to their solution which interests us.[1]

The so-called sweetness of the problems provided its own momentum. Even if equivalent studies were being conducted abroad, the process of performing experiments was, in itself, an attractive proposition to the Japanese, who were still "catching up" with the West and who were used to following the example of others.

In the previous chapter, we saw the conditions that led to the growth of physics in Japan. It is significant that one of Japan's most important physicists would never occupy an academic post in a university, but, instead, be attached to an autonomous research institution. The physicist was Nishina and the place was Riken, also known as the

Institute of Physical and Chemical Research. When we consider Nishina's achievements, it is not so much for specific scientific contributions that we remember him.[2] His claim to fame rests in the achievement of building machines (cyclotrons) that others (notably Ernest Lawrence) had invented; and in his ability to mobilize the necessary resources and personnel. This chapter outlines how he went about doing this, especially during World War II.

By first tracing his family background and education, formative influences of the type discussed in the previous chapter will be seen to have shaped Nishina's career path and choices. His decision to build a cyclotron marked the beginning of the Americanization of Japanese physics by which Berkeley physics came to Tokyo at a time when Japan was on the verge of a war and would ultimately go to war with the United States. The mobilization of science is outlined in the second section of this chapter. How did physicists respond to the call to assist in the war effort? This is followed by an account of how it directly affected Nishina in the atomic bomb project. The final section tells how it came to an end.

The relevance of this chapter is in how Nishina, a physicist with no imperial university affiliation, was able to overcome the odds and convince the Army and other funding bodies to part with large sums of money. How did physicists reconcile working on such a project? How physicists have gone about obtaining funding, their motivations, and what they have argued for is what the following chapters are all about.

I   YOSHIO NISHINA: FAMILY BACKGROUND, EDUCATION, AND RESEARCH

Like many physicists of his day, Yoshio Nishina came from a privileged background. He was born on December 6, 1890, into a family with a tradition of public service, in the village of Satoshō-chō in the district of Asakuchi, located in the southwestern part of Okayama prefecture. Yoshio was the fourth son and one of the nine children of Arimasa and Tsune Nishina. Nishina's ancestors had, in pre-Meiji days, held the position of "*daikan*" ("local official") and they were a wealthy, landowning family. The main task of a *daikan* appointed by feudal domains was of a financial nature: the levying of taxes.[3] Yoshio's grandfather, Arimoto, had looked after the finances of the Ikeda clan, and had been responsible for issuing a type of local paper money. After the Meiji Restoration, when such local monies were abolished, it is said that Arimoto sacrificed his own wealth in order to call back the notes that he had issued. In the Meiji period, the position of *daikan* was abolished and the Nishinas turned to agriculture and salt production. The income from such ventures was not great and it is likely that the Nishinas experienced considerable hardship. Nevertheless, the status of the family remained high and Yoshio grew up in an environment of privilege, the family home being the *daikan* residence of grandfather Arimoto.

After eight years of elementary school, Nishina successfully gained admission to the Okayama Middle School in 1905—the year in which the Russo-Japanese War came to a triumphant end for the Japanese. The school was actually housed in the former castle in Okayama city. Because of the considerable distance from home, Nishina boarded at the school for five years. He went on to attend the Sixth Senior High School in the same city, graduating in July 1914 at another turning point in Japanese history. Unfortunately, Nishina came down with pleurisy and missed a year of schooling. (He would contract it again while at university.) Nishina, thus, spent a total of nine years in Okayama away from his family.[4]

The Nishinas were a talented family. Yoshio's second eldest brother, Tōhei, had left home to live in Tokyo, where he invented such things as processes for dried foodstuffs and fireproof paint. In September 1914, Nishina entered the Department of Electrical Engineering in the School of Engineering of Tokyo Imperial University. Yoshio would later live at the home of his brother while studying there. Nishina graduated at the top of his class in electrical engineering and contemplated joining the Shibaura Engineering Works, which specialized in the manufacture of electric machinery and apparatus. The company was the first private engineering firm in Japan. It was established in 1875 by Hisashige Tanaka, and later became part of the Mitsui *zaibatsu* (family financial combine), and the leading enterprise of its type.[5] Nishina felt, however, that the field of electrical engineering lacked potential for much development and considered that electrochemistry might be more rewarding. He was likely to have been influenced by the young professor Kōtarō Kujirai of the Department of Electrical Engineering. Kujirai had recently returned from a period of study abroad and had just been appointed as a research fellow at Riken. After graduation in July 1918, Nishina entered the newly established Riken as a research student in the Kujirai laboratory.

Soon afterward, Nishina entered graduate school. Rather than studying chemistry, Nishina chose physics and math, possibly on Hantarō Nagaoka's advice. But Nishina lacked sufficient physics to embark immediately on research in the area. Nagaoka was head of the physics division at Riken and, in the future, would serve as one of Nishina's patrons. Nishina was able to attend university lectures given by Nagaoka and gained a strong interest in physics. He spent almost three years in graduate school auditing physics lectures he had missed as an undergraduate. Like other students, Nishina was also affiliated with Riken. Before long, he was appointed a researcher there. The blurring of the boundary between science and engineering in Nishina's education was unusual, but can also be seen in the careers of other prominent Japanese scientists.[6]

In 1921, Nishina went to Europe as a Riken "*kaigai ryūgakusei*" ("overseas student"). His first port of call was England where he studied with Ernest Rutherford at the Cavendish Laboratory, Cambridge University, and met the likes of James Chadwick, Henry Smyth, and Peter Kapitza. In 1913, Rutherford and Geiger had devised an electrical method by which to count individual atomic particles.[7] Nishina had opportunities to use the Geiger counters and carry out research into the distribution of recoil electrons in the air in the Compton scattering of x-rays. Nishina remained at the Cavendish until August 1922. In November, he traveled to Germany, studying at Göttingen University until March 1923. He was able to attend lectures given by scientists such as the physicist Max Born and the mathematician David Hilbert. High inflation in Germany, at the time, made living expensive, and Nishina went on to Denmark for a more affordable lifestyle. He would spend about five years (April 10, 1923–September 30, 1928) with Niels Bohr in Copenhagen, and a total of seven years in Europe. This was at a formative time in the development of quantum mechanics, when scientists from throughout the world gathered at Bohr's Institute of Theoretical Physics in Copenhagen to discuss theory and conduct related experiments.[8]

During Nishina's own lifetime, considerable developments in physics had occurred. Shortly after Nishina was born, W.C. Röntgen had shown that x-rays could penetrate thick matter, the absorption being approximately proportional to the atomic weight of the matter penetrated. Many years later, Henry G.J. Moseley showed the connection between x-ray spectra of the elements to the electronic structure of the atoms of which they were composed. Bohr's theory of atomic structure was supported by absorption spectra from experiments.[9] In 1923, Nishina, along with Dirk Coster and Sven Werner, used x-ray absorption spectra to investigate the relationship between atomic structure and L-absorption spectra of elements of atomic number from 57 to 72 (lanthanum to hafnium). At Copenhagen, Nishina would coauthor a theoretical paper with Oskar Klein on the rate and angular distribution of Compton scattering. Much of Nishina's work was collaborative but he did write an individual paper related to this work for *The Philosophical Magazine*, which was published in 1925.[10]

Nishina was not the only young Japanese scientist in Copenhagen. A number of other young Japanese visited Bohr's Institute, including Toshio Takamine (1921, 1925), Kenjirō Kimura (Tokyo, 1925–1927), Yoshikatsu Sugiura (1925–1927), Takeo Hori (1926–1927), Shin-ichi Aoyama (Tohoku University, 1926–1927), and Mitsuharu Fukuda (1928–1929).[11] Nishina collaborated with Kimura and Aoyama in a study of the variation of x-ray absorption spectra in relation to the chemical combination of elements.[12]

Nishina had initially relied on funds provided for overseas study by Riken and also on a Tokyo Imperial University scholarship. When such funds ran out in 1923, Nishina received assistance throughout the period 1924–1927 from a Danish government scientific fund called the Rask-Ørsted Foundation for the promotion of international cooperation in science. While the Foundation funding was meant to support visiting scientists for only one or two years, Bohr could argue a case for extended funding for scientists such as George de Hevesy and Nishina.[13] But this could not last forever. From August to October 1927, Nishina visited Paris,[14] after which he worked on theoretical research with Wolfgang Pauli at Hamburg University (November, 1927–February, 1928). During this time, he had the opportunity to collaborate with I.I. Rabi on a joint paper. In 1928, Nishina returned to Copenhagen and, with Klein, carried out theoretical research into the Compton scattering of penetrating x-rays (gamma rays) by a free electron according to Paul Dirac's new theory of relativistic quantum mechanics. This would lead to the Klein–Nishina formula. He left Copenhagen at the end of September to return to Japan. On his way home, Nishina visited a number of American universities in New York, Chicago, and San Francisco, arriving in Japan on December 21 after a period of over seven years overseas.[15] It is indicative of Nishina's resourcefulness and sheer talent that he was able to survive for such a long time on limited funds. It was fortunate that others around him, such as Bohr and Nagaoka, recognized that talent and supported his research. Despite Nishina's absence from Japan, he had the opportunity to meet with Nagaoka during the latter's visits to Europe.[16]

Upon his return, Nishina found it difficult to obtain a university position in physics because of his unconventional background in engineering and long absence from the scientific community. Bohr had written Nishina a long letter of recommendation to assist him in finding employment, but Nishina knew that such a document, despite the signature, would not be of much use in Japan apart from reminding "me of the years I spent in your Institute and it will ever encourage me in time of depression."[17] Instead of a university appointment, Nishina was given a position in the Hantarō Nagaoka laboratory at Riken and assigned the research topic of "quantum mechanics and its applications." Fresh and enthusiastic from his time overseas, Nishina found the time to marry Mie Nawa, the younger sister of close university friend Takeshi Nawa, on February 23, 1929, and also to set about organizing a visit to Japan by Heisenberg and Dirac. He acted as their interpreter when they actually arrived in September 1929 for a visit of three weeks. The visit was sponsored by the Keimeikai Foundation, an organization for the promotion of scholarship to which Nagaoka had applied.[18] Visits by such leaders in the field of physics undoubtedly enhanced Nishina's reputation. The following year Nishina obtained his Doctor of Science degree.[19]

At the beginning of 1930, Japan returned to the gold standard, effectively depreciating Japanese currency to less than one half of its normal value and making it very difficult for Japanese to go abroad.[20] Nishina did not have much option but to stay in Japan and hope for a senior research appointment. In May 1931, he gave a series of intensive lectures on quantum mechanics at Kyoto University to which a number of budding physicists, including Hideki Yukawa, Sin-itirō Tomonaga, Shōichi Sakata, and Minoru Kobayashi, would go. Shortly afterward, on July 1, Nishina was appointed a chief researcher at Riken with strong backing from Nagaoka. Nishina was the youngest person to hold this position, which came complete (in 1932) with his own laboratory. The year 1931, which was also the year of the Manchurian Incident, would signal the beginning of a period in which the military would become the dominant power group in Japan and nationalistic feeling would become widespread. It was thus a year of coming-of-age for Nishina as a scientific entrepreneur and, in some respects, "maturing" of Japan as a nation. Nishina's activities as an organizer and facilitator tended to dominate all else. It has been noted that from this time on, Nishina would never be the single author of a scientific paper again. Rejection of his credentials by university-based academia facilitated his move in this direction. The political environment was also conducive to his mobilizing of funds for research projects.

Early members of Nishina's laboratory were Hantarō Nagaoka's son Ryōkichi Sagane, Masa Takeuchi, and Sin-itirō Tomonaga.[21] The first two had joined the Nagaoka lab in 1931, working with Nishina and using Geiger-Müller counters. The Geiger-Müller counter, also known as the "discharge ionization chamber," had only been perfected a few years earlier in 1928. An independent "Nishina laboratory" existed from 1932.[22] Nishina described the lab to Bohr as his "room of my own" for experimental work on the atomic nucleus and cosmic rays. Given the sharp decline in the value of the yen, he expressed concern over how he might fund such research.[23] Despite such fears, Sagane and Takeuchi were able to set up a Wilson cloud chamber[24] to observe trajectories of particles, Geiger-Müller counters, and, in 1933, a Van de Graaff accelerator. The Van de Graaff apparatus (and also the Cockcroft–Walton machine, which would be built later) was a means by which to obtain high-speed ions using high voltages. Neutrons had yet to be discovered and it was thought that only charged particles (such as protons or alpha particles) could be used to investigate the nucleus. These particles would have to be speeded up if they were to overcome the repulsion of the nucleus.

While Nishina can be credited with helping to bring the "Copenhagen spirit" of Bohr's Institute to Japan, his lab transplanted the American style of doing particle physics in Japan as well. Nishina had quickly realized that nuclear physics could no longer be done on small-scale equipment using sources such as polonium (alpha particles) and radium. Machines such as the cyclotron,[25] invented in 1930 by Ernest O. Lawrence at the University of California, Berkeley, were the "big science" of the 1930s, and big science required big funding. The cyclotron involved a new procedure by which ions were accelerated to high speeds in a number of steps that did not involve a huge voltage. The ion would travel a spiral path, which would keep the accelerator down to a manageable size.[26] Around this time, discoveries were coming hot and fast: the discovery by C.D. Anderson of the positron, which Dirac had predicted (1932); H. Urey's discovery of deuterium (1932); the neutron by James Chadwick (1932); and artificial radioactivity by I. Curie and F. Joliot (1934). Such discoveries led Nishina, in 1933, to conduct a study with Sin-itirō Tomonaga into the pair creation of positive and negative electrons by gamma rays, according to Dirac's theory. Also, shortly after the discovery of artificial radioactivity, Nishina set about measuring the energy spectrum of positrons emitted when radio-phosphorus decays, using a Wilson cloud chamber. Neutrons soon became known as effective producers of radioactive isotopes, which could be used as tracers.

On September 18, 1931, an explosion occurred on the South Manchurian Railway line just outside Mukden. The incident was blamed on Chinese troops and used by the Kwantung Army as a pretext for occupying Mukden. In February 1932, not long after the Manchurian Incident, the puppet state of Manchukuo was established. Amid international protests, Japan withdrew from the League of Nations. Japanese troops gradually made further inroads into China, the hostilities leading to a full-scale war in 1937. Nishina, writing to Bohr's secretary Betty Schultz in 1933, wrote of how international events at first made little immediate impact on physicists at Riken:

> We do not take it so serious. You asked if we have been to war. No, none of us has the least chance of going to war. We are all doing physics as peacefully as ever.[27]

Nishina repeated similar sentiments to Hevesy that year:

> In Japan it is the time of reaction too [as compared to Germany] and I suppose Japan is not very popular among most of [sic] foreign countries on account of the Manchurian affairs. We cannot help it, because that seems to be a national movement. For myself I am happy to be able to do science without being troubled with political affairs. Only thing I feel very troubled [about] is the depreciation of our currency. It is nearly impossible to get any experimental apparatus from abroad, we have to get them made in Japan. That takes sometimes very much time.[28]

It was opportune for Nishina that in January 1934, the Japan Society for the Promotion of Science No. 10 Committee was formed in order to advise on funding and deliberate on policy for studies of cosmic rays and the atomic nucleus. With the exciting discoveries occurring in nuclear physics, a Japanese facility for large-scale experiments in nuclear physics was deemed necessary, and, given the shortage of funds elsewhere, it was decided to create a laboratory in 1935 at Riken, which would be jointly run by Shōji Nishikawa and Nishina. The principal items of equipment would be a Cockcroft–Walton apparatus[29] and a cyclotron. A cyclotron would produce powerful neutron sources, which, in turn, could make many new isotopes of elements. The establishment of the laboratory was a major political feat as nothing of its kind existed in Japan at the time. Sin-itirō Tomonaga viewed Nishina's campaign for the establishment of the laboratory in highly heroic terms:

> A laboratory for research in new physics, whether it be nuclear physics or cosmic ray physics, is a laboratory of a very large scale which is furnished with a monster of a crane above our head and a gigantic

high-tension [high-voltage] apparatus weighing as much as a few hundred tons, and which has an electric source and machine shop attached to it besides [*sic* in addition]—in short, it looks like a factory, so to speak. How laborious a task it must have been, and how painstakingly he must have worked, to enlighten the people and make them understand that a [*sic*] research in pure physics needed such a big "factory." But at last his earnest campaign of enlightenment succeeded, and a nuclear research laboratory was made in the Institute of Physical and Chemical Research. . . . The result of his campaign was not limited to that only, but some universities came to have their own "factories" of nuclear physics furnished with such majestic apparatuses as cyclotrons and so forth, for people at large were persuaded by him and began at last to perceive the importance of the study of [the] atomic nucleus.[30]

Visits to Riken by dignitaries such as Prince Chichibu on December 4, 1934 (and again on March 16, 1937 to inspect the nuclear-physics and cosmic-ray laboratory as well as the small cyclotron) were part of a public relations strategy, which was deemed necessary for fundraising. Despite Nishina's dislike of journalists, he knew that publicity through newspapers, magazines, and open days would help persuade the business community to part with some money. One of Nishina's sons, Yuichirō Nishina, fondly remembers the stunt in which a staff member drank salt water containing a sodium isotope. A Geiger counter was used to trace the movement of the radioactivity through the person's body. This was around the time of one of Prince Chichibu's visits to Riken.[31] This talent for public relations was also shared by Ernest Lawrence who, on his lecture circuit, would serve radio-sodium cocktails to a volunteer and later trace the progress of the isotope through that person's bloodstream.[32]

The Nishina laboratory took charge of the cyclotron project. In mid-1935, Nishina arranged with Lawrence for Sagane to go to Berkeley and the San Francisco area, the center of accelerator building. Sagane virtually served a short apprenticeship in the fine art of cyclotron making. He arrived at Lawrence's laboratory in September 1935 and was joined for a brief period later that year by Tameichi Yasaki, another Nishina-lab researcher. Nishina was now busier than ever.[33] In 1936, the plans for a "small," now 37-inch (rather than 27-inch as originally planned), 23-ton machine were finalized with the help of information provided by Sagane and others in Berkeley and construction was completed in April 1937 with donations from the Tokyo Electric Light Company (the president of which was Yasaki's wife's uncle) and the Mitsui Foundation.[34] With this machine, new research was conducted in nuclear physics, radiobiology, and in the use of radioactive tracers.

One is tempted to canvas the question of whether Japan of the 1930s and the increased opportunities for research funding, or Nishina's personal powers of persuasion, were more influential in securing the transfer of cyclotron technology from Berkeley to Tokyo before other nations joined the Berkeley big science. Nishina cultivated private and public patrons such as the senior physicist Nagaoka and Riken's Ōkōchi. Nagaoka belonged to the Science Deliberative Council (Kagaku Kyōgikai), which was first established in 1921 to improve liaison between scientists and the military. This Council became the Society for Science for National Defense (Kokubō Kagaku Kyōkai) in 1934, with Nagaoka, at one stage, being a director. Nagaoka was also Chairman of the National Research Council Committee for Radio Research, and his roles as Division Head and later Chairman of Directors of the Japan Society for the Promotion of Science meant that he would become heavily involved with the military on matters of science.[35]

In autumn 1934, the Ministry of War criticized the individualism of the Japanese in a well-known pamphlet entitled "Call for the True Meaning of National Defense and its Strengthening" ("Kokubō no hongi to sono kyōka no teishō"):

> The present economic system has been developed on the basis of individualism. For this reason economic activities tend to serve only individual interests and fancies, and do not always harmonize with the general interests of the State.
>
> It is desirable that the people should abandon their individualistic economic conceptions; instead they should recognize the importance of a collective economy; they should work towards the creation of an economic system which will rapidly realize the Empire's ideal.
>
> The State should rigidly control the entire national economy.[36]

Despite such calls for mobilization, the Nishina laboratory went about its "individualistic" way and began two large projects on cosmic rays: the measurement of cosmic ray intensities in the Shimizu tunnel (August 1936), and the study of cosmic ray particles by a large Wilson cloud chamber (1937). The Nishina lab measured the mass of the meson in cosmic rays by using the cloud chamber. In addition, Nishina, backed by Nagaoka, was instrumental in having Bohr visit Japan in April and May 1937, coincidentally, just as the small cyclotron began operation. Nagaoka took care of funding by approaching *zaibatsu* such as Mitsui and Mitsubishi to provide 20,000 yen (the equivalent of around £1000) to cover Bohr's expenses.[37] The completion of the small cyclotron in 1937 boosted Nishina's confidence, especially as it worked better than expected. Before the small cyclotron even got going, Nishina had plans of building an even larger (40–50 inch) one, which would produce a beam of around 25 MV and had started fundraising for it in 1936. Lawrence suggested that to keep costs down, the 200-ton electromagnet be built in the United States. Nishina agreed and in 1938, the magnet and other materials arrived in Japan. The physicists busied themselves with the construction of the cyclotron and, as a result, the small cyclotron came to be used mainly for biological and medical research. The Press called Nishina's factory "The Magic Laboratory":

> The Magic Laboratory, recently completed in the Institute of Physical and Chemical Research, has started the operation of its machine on April 6, and the unusual artificial radium has begun to run out from it for the first time in Japan. Part of the Nuclear Laboratory has been completed, and on April 6 [1937] began to produce the world's treasure "radium" which costs 200,000 yen per gram. . . . On April 5, the Japan Society for the Promotion of Scientific Research decided to grant 110,000 yen to the laboratory and the latter therefore will begin a series of large scale experiments. The world leading figure in atomic physics and quantum theory, Dr. Bohr, will arrive in Japan on April 15, and will stay at the Institute for a while to do research with its staff.[38]

For the journalists, the coincidental timing of Bohr's visit vouched for the credibility of Nishina's projects.

The subordination of the individual to the group that the Ministry of War called for was also a characteristic of Berkeley physics, which was spreading throughout the United States and to the rest of the world. After studying at Berkeley, Sagane toured other U.S. labs and the Cavendish Laboratory, confirming for himself the preeminence of the Berkeley style of physics. Writing to Lawrence from the Cavendish in November 1936, Sagane mentioned how

> I was rather disappointed and also astonished of [sic] their rather poor apparatus. . . . It seems to me so far as the experimental techniques are concerned, America has surpassed very far the England.[39]

With help from Lawrence and Sagane, Nishina became Japan's expert on things nuclear. It is thus not surprising that the military would turn to Nishina when canvasing the idea of an atomic bomb. It has been suggested by physicists in Nishina's lab at the time that his eventual acquiescence to becoming involved in an atomic bomb project was in order to divert military funds to complete the large, eventually 60-inch, cyclotron.[40] Despite outside funding and help from Lawrence, the construction of the cyclotron had not gone smoothly. Around 1937, Nishina became concerned that if the war with China was prolonged, scientific research would be starved of funds and progress on the cyclotron would be hampered. Obtaining parts for the cyclotron from overseas became more difficult as the government regulated imports to reduce foreign debt and maintain the rate of exchange.[41] Nishina adopted cost-saving measures, such as purchasing heavy water (deuterium oxide) directly from abroad. George de Hevesy assisted by placing an order with a firm in Norway in 1937.[42]

Nishina was decidedly multidisciplinary. His training in electrical engineering and research in physics prepared him well for the construction of a cyclotron and so-called factory physics. His keen sense of the needs of the wider society matched well with the potential of the cyclotron for use in medicine and biology. It has been suggested that these possible applications appealed to the funding bodies. According to Sin-itirō Tomonaga, one of the physicists in Nishina's lab, it was notorious for running over budget, but the Institute would invariably bail the lab out of its debts.[43]

The discovery of nuclear fission by Otto Hahn, Fritz Strassmann, Otto Frisch, and Lise Meitner in Europe at the end of 1938 prompted considerable interest among scientists in Japan. Hahn was quite envious that the Japanese possessed a cyclotron. It appears that Hahn complained to Lise Meitner about the slowness of progress in their work due to the lack of sufficiently strong sources.

> The trials are proceeding slowly. . . . In the English Nature Journal five Japanese [Nishina et al.] report on the formation of UY [Uranium Y] out of thorium and neutrons. They have a cyclotron. . . . If we had stronger radiation sources we would get along faster.[44]

Physicists such as Sin-itirō Tomonaga and Hideki Yukawa were more concerned with their own theoretical research than with the applications of fission, despite the opportunities to study the latter. Tomonaga was actually studying at Leipzig University with Werner Heisenberg (later an important figure in the German atomic bomb project) between 1937 and 1939. In Tomonaga's diary, there are entries that mention seminars on nuclear fission given by Strassman and Carl F. von Weizsäcker in early 1939.[45]

The travel diary of Yukawa during his trip to Europe and the United States in 1939 for the Eighth Solvay Conference in Brussels, mentions visits to the Institute for Theoretical Physics in Leipzig and to Columbia University to see Enrico Fermi.[46] Despite numerous chances to become involved in the application of nuclear fission, both physicists chose to concentrate on meson theory. In 1949, Yukawa became the first Japanese Nobel laureate for his work in theoretical physics, part of which he carried out during the war. Tomonaga received the Nobel Prize for his contribution to elementary particle physics in 1965, again for work that evolved partly during the war.

In 1939, hearing of the splitting of the uranium nucleus, physicists used the small cyclotron to confirm the European discovery by conducting experiments on the nuclear fission of thorium by neutrons, and the fission products of uranium by fast neutrons. Unlike Yukawa and Tomonaga, Nishina launched himself into fission experimentation, writing a number of papers on the subject in 1940 and 1941. The laboratory would in time look at the enrichment of uranium 235 and the use of chain reactions for the production of atomic energy. Although Yukawa and Tomonaga enjoyed a close relationship with their teacher, there was no compulsion to become actively involved in atomic bomb projects. In 1939, Yukawa was appointed professor of theoretical physics at Kyoto University. Sakata moved from Osaka University to Kyoto to join Yukawa, along with Yasutaka Tanikawa. Yukawa became interested in establishing a reformulation of the relativistic quantum field theory, a problem he had grappled with years earlier. He hoped to construct a comprehensive theory of elementary particles with no divergence problems.[47] In 1940, Sakata and Tanikawa predicted the gamma decay of a neutral meson and in 1942, the Kyoto group of Seitarō Nakamura, Takeshi Inoue, Tanikawa, and Sakata proposed the two-meson theory. Sakata was appointed professor that year to the newly established science faculty at Nagoya University, and Inoue and Tanikawa subsequently joined him there.[48]

Upon his return to Japan, Tomonaga, who had studied nuclear physics and quantum field theory under Heisenberg, directed his attention to the meson theory. In 1940, he developed the intermediate coupling theory in order to clarify the structure of the meson cloud around the nucleon, and published it the following year.[49] In 1941, Tomonaga was appointed professor of physics at the Tokyo University of Science and Literature, but still retained links with Riken. In 1942, he first proposed his "super-many time theory," which was identical to the covariant field theory later developed by Julian Schwinger. Applied to quantum field theory, it provided a powerful framework for renormalization theory.[50]

The "internationalism" of science was overtaken by national loyalties with the beginning of World War II. This was made clear to Nishina in August 1940 when physicists from his laboratory were prevented from inspecting the Lawrence laboratory at Berkeley, to which they had previously been welcome. As Lawrence diplomatically put it to Nishina,

> The reasons for this sudden restriction have to do with the fact that in many of the laboratories of the university there is at the present time marked overcrowding, and also there is a certain amount of work in progress of a confidential character.[51]

With the outbreak of the Pacific War, Japan's scientific links were cut off from all countries save Germany and Italy.[52] What technical exchange occurred between Japan and Germany was strictly on a reciprocal benefit basis. Military intelligence, political support, or raw materials were given in return for German technical ideas. But as Admiral Raeder of the German Navy stated in 1941,

> The conclusion of the Three Powers' Pact as such does not bind us in any way to reciprocal gestures. The measure of response to Japan, which thinks it can demand that Germany sell its birth-right under the motto of it being "essential for the fulfilment of the Three Powers' Pact" is solely dependent on the state of military–political interests at any given time.[53]

From that year onward, the Japanese sent "wish-lists" of desired weapons to the Germans, going so far as to request an entire aircraft factory. Hitler and Albert Speer, apparently, approved this and other requests on the promise of rubber, tungsten, and victuals, but neither side was able to fulfill their part of the bargain.[54]

Despite the lack of internationalism in wartime physics, Tomonaga's field theory formulations and Sakata's two-meson theory were formed.[55] The activities of the Meson Club were undoubtedly a factor contributing to their success. The term "Meson Club" is loosely used to refer to a series of symposia and informal discussions on meson theory, which occurred from 1937 to 1944. The twice-yearly meetings at Riken and the annual meeting of the Physico-Mathematical Society often provided a venue for the Meson Club discussions.[56]

Although the meson theory had, at first, enjoyed a degree of success, its shortcomings became apparent as studies of cosmic rays and nuclear research advanced. Taketani suggested a basic plan of dividing activity into three parts: phenomenological, substantialistic, and essentialistic.[57] In 1943, Tomonaga, who still remained in charge of theoretical research in Nishina's laboratory, divided research broadly along these lines. Hidehiko Tamaki and Taketani studied the appropriateness of the meson model in view of experimental data. Gentarō Araki and Sakata looked into applications of quantum theory and their limitations, whereas Tomonaga maintained responsibility for mathematical methods.[58] Araki produced strong evidence for a pseudo-scalar meson.[59] Sakata developed his two-meson theory, which assumed that the meson responsible for nuclear forces was different from the meson observed in cosmic rays. Tomonaga published his renormalization theory, which made quantum electrodynamics consistent with the special theory of relativity.[60]

Particularly after the first two years of the Pacific War, however, scientists were successively pressured into working on military research. Physicists were not entirely free to conduct pure research. Inoue was conscripted into the army. Taketani's research and health suffered as a result of imprisonment.[61] After 1943, Tomonaga temporarily stopped elementary particle physics research and became involved in electronics, after a request from the Naval Technical Research Institute. He conducted research at the Shimada Laboratory in Shizuoka prefecture from 1944 till the end of the war.[62]

Midway through the construction of the large 60-inch cyclotron, there were changes in the plans and, as a result, there were delays. Construction ended in February 1943, after which various adjustments had to be made. In January 1944, the laboratory succeeded in obtaining deuterium ions at 16 million volts. Throughout 1944, Nishina had problems in obtaining powerful vacuum tubes for his cyclotron.[63] It was testimony to Nishina's political and entrepreneurial skills that such research could be continued even into the Pacific War. It is surprising that only part of the Nishina laboratory was mobilized for military research.

II MOBILIZATION OF SCIENCE

Till late 1937, the Japan Society for the Promotion of Science (Nihon Gakujutsu Shinkōkai) was the principal force behind the growing bureaucratization of science.[64] Until then, scientists had been regarded somewhat suspiciously as being potentially subversive and unpatriotic in their "objectivity."[65] Even foreign observers agreed that as many of Japan's top scientists had been educated in the United States or Europe, the military viewed them as foreign sympathizers. Furthermore, having traveled abroad would have enabled them to see

Japan's policies and capabilities in better relation to that of other countries.[66]

A full-scale war with China broke out at Shanghai on August 13, 1937. In October, a Science Division (Kagakuka) was created within the newly established Cabinet Planning Board (Kikakuin) to integrate research activities into the wartime economy. However, following major military setbacks, attempts were made in 1938 to mobilize Japanese science for the war effort. With the promulgation of the National Mobilization Law, the Japanese government became directly involved in attempts to centralize science. The government immediately established the Science Council (Kagaku Shingikai) within the Cabinet, virtually incorporating the Society for the Promotion of Science. Given that the role of Cabinet's Science Division had now changed to the mobilization of research, it was upgraded to Science Department (Kagakubu). These organizations would later be incorporated into the powerful Agency of Science and Technology. The Ministry of Education, meanwhile, established its own Association for the Advancement and Investigation of Science (Kagaku Shinkō Chōsakai), the forerunner of the Science and Technology Council. Not to be outdone by Cabinet, the Association recommended the establishment of a Science Division (Kagakuka) within the Ministry of Education. The Division controlled university and college research, and the funding of other research organizations of the Ministry of Education. Sectionalism within government, thus, led to a disastrous, two-tier approach to the mobilization of science.[67]

Attempts to mobilize scientists took two general directions. The first was the establishment of research laboratories, funded by the government and often attached to universities. The second was the creation of organizations to liaise with those working in research.[68] The Japan Federation of Science and Technology Organizations (Zen-Nippon Kagaku-Gijutsu-sha Dantai Rengō) was established on August 2, 1940 and assumed much of the responsibility for implementing the policy. The Federation was linked with the Ministry of Education's Science Division through the National Research Council. The Federation attempted to impose a structure aimed at breaking down academic cliques and preventing overlap in research. Its efforts toward rapid rationalization of science, expectedly, met with a great deal of opposition from the scientific community.[69]

The Cabinet Planning Board's Science Department set up an auxiliary organization to do this as well. On December 8, 1940 the Science Mobilization Association (Kagaku Dōin Kyōkai) was created with the aim of bringing together scientists of different fields and affiliations. In order to realize this, the Association, in collaboration with the Navy, conducted a survey of the research interests of all scientists. In October 1941, survey cards were distributed and a plan of strategy was later formed, based on the findings. It was envisaged that the 32 different areas of research would be organized into 262 subsections. Several thousand research units would be formed within these subsections, according to topic, and these would, in turn, fit into a pyramidal structure or hierarchy.[70]

The war effort provided the opportunity for a concerted attack on the feudalistic manner in which research had been conducted in Japan. Scientists were blamed for creating a gulf between basic and applied research, and of encouraging sectionalism. In order to overcome these problems, a "General Plan for the Establishment of a New Scientific and Technological Structure" ("Kagaku Gijutsu Shintaisei Kakuritsu Yōkō") was passed by Cabinet on May 27, 1941.[71]

In January 1942, the Cabinet Science Department was expanded to become the Agency of Science and Technology (Gijutsuin), a Japanese version of the U.S. Office of Scientific Research and Development (OSRD). The Agency was founded to oversee the general administration of the new structure, and to pay special attention to aeronautical engineering. Responsibility for the Central Aeronautic Laboratory was transferred from the Ministry of Communications to this Agency.[72] Later that year, the Science and Technology Council (Kagaku Gijutsu Shingikai) was established by the government to deliberate on policy and to provide central support for mobilization. However, despite their efforts, the mobilization of science reached a period of stagnation. This was partly a result of territorial struggle amongst government ministries. The Ministry of Education, not wishing to lose ground, expanded its former Science Division to Science Department in 1942. This inter-ministry antagonism extended to the Army and Navy.[73]

Japan was buoyed by its easy victories early in the Pacific War, and there was a tendency for inaction on matters such as unifying the mobilization of science. What mobilization efforts there were, were mainly aimed at applied research and a policy was adopted of not interfering with basic research conducted at universities.[74] Physicists did, at least, pay some heed to the increase in nationalistic feeling in their public utterances. Hideki Yukawa had the following *tanka* poem published in 1942:

Kuni ni sasagu inochi nao arite
Kyō mo yuku hitosuji no michi
Kagiri naki michi

Still possessing a life which can be devoted to the nation,
I continue to proceed earnestly along the road, an endless road.[75]

The poem emphasizes the duality of a life devoted to the nation and to the path of science. Science and the State are represented as but two sides of the one. Yukawa's poem is hardly surprising as government rhetoric had long been calling for the denial of the self in favor of the nation, and

> with few exceptions writers . . . conformed to the policies of the military, stifling whatever doubts they may occasionally have felt. Like all other Japanese, they rejoiced in military victories and lamented defeats. . . . Only a few writers, mainly established authors, could afford the luxury of remaining aloof from the war effort; the others had no choice but to compose works that demonstrated their patriotism and encouraged fellow Japanese to fight even harder.[76]

Nagaoka was not shy about engaging in such rhetoric and was a central figure in the mobilization of Japanese science. It appears that he enjoyed a special relationship with the Army and Navy. In 1940, he became a consultant to the Army Technical Headquarters and councilor of the Army Ordnance Department. Nagaoka belonged to many military and Agency of Science and Technology committees. He held membership in the Navy's own Science and Technology Council (Kaigun Kagaku Gijutsu Shingikai) (1942) and the Army and Navy Committee for Radio Technology (Rikukaigun Denpa Gijutsu Iinkai) (1943). He was also a councilor of the Agency of Science and Technology (1944). Around 1943–1944, he was employed as a part-time consultant by the Naval Technical Research Institute.[77]

With the worsening of the war situation, the mobilization of science reached new levels. In August 1943, the "Emergency Plan for Scientific Research" ("Kagaku Kenkyū Kinkyū Seibi Hōsaku Yōkō") was conceived. Under this plan, the National Research Council (Gakujutsu Kenkyū Kaigi) created a Research Mobilization Committee (Kenkyū Dōin Iinkai) with liaison committees in each university to regulate research.[78]

In October 1943, Cabinet decided to reorganize the overall mobilization system again. A Research Mobilization Council (Kenkyū Dōin Kaigi) was established to decide important areas of research, and to take appropriate action for their development. Suitable personnel would be recruited for military research, and sufficient funds and materials supplied for their needs. The government's "Integrated Plan for the Mobilization of Science and Technology" ("Kagaku Gijutsu

Dōin Sōgō Hōsaku") gave priority to aeronautic research and the development of new scientific weapons. Research was to satisfy and comply with military requests. Other features of the plan were: a technical guidance system for production; a new system of registration for scientists; and a channel to be created through which original ideas could be received from the public.[79]

In November 1943, the General Affairs Section of the Agency of Science and Technology appealed to the nation for creative input in the form of suggestions for devices and weapons. A committee was convened to examine the inventions and ideas that were submitted. The registration of scientists and technologists was conducted from March to May 1944 by the Ministry of Welfare. Cabinet decided to give priority to the promotion of scientific and technological research in budget considerations, and researchers in the basic sciences began to participate in military research in increasing numbers from this time. In March 1944, a group of 66 researchers was specially appointed to conduct such work. Of the participants, 43 were full or associate professors from universities; 11 came from government agencies; and the remainder from the private sector.[80]

In the closing stages of the War, fresh measures were taken in a desperate attempt to muster science and technology to avert defeat. On August 24, 1944, the Ministry of Education, in consultation with the Ministry of Welfare and the military, decided to mobilize students for wartime research. One thousand science students, in at least their second year of study, were selected from technical colleges and universities and deployed to assist in research.[81] On September 5, 1944, the Standing Committee for Army and Naval Technology (Riku Kaigun Gijutsu Unyō Iinkai) was created in a last-ditch attempt by the government for centralized control of new weapons research and production. In January 1945, there was a structural reform of the National Research Council, aimed at promoting better liaison between the academic world, the military, and government agencies.[82]

This chronological description of official measures taken by the Japanese government tends to give the picture of a highly organized effort. However, the mobilization of science did, in effect, meet with only limited success, and when compared to efforts in the United States, was a downright failure. Nevertheless, it is important to trace the activities of Japanese physicists during this period and examine how they responded to the call for service to the nation. Given Nishina's considerable success in attracting research funds, it is not surprising that his entrepreneurial skills would be put to good use during the Pacific War.

III THE ATOMIC BOMB PROJECT

At the beginning of the Pacific War, the Nishina lab consisted of a number of groups: nuclear physics groups centered around the cyclotron, cosmic ray specialists, a theory group, and a group studying the effects of radiation on living organisms. The first category included Gorō Miyamoto of Tokyo Imperial University. Ryōkichi Sagane and Yatarō Sekido were among those involved in cosmic-ray research. Sin-itirō Tomonaga, Minoru Kobayashi, Shōichi Sakata, Hidehiko Tamaki, Mituo Taketani, and Nobuyuki Fukuda formed part of the theory group.[83]

The 1940s saw a transition to military-related research for Nishina's laboratory and physicists in general. In 1940, Nishina and his laboratory studied the influence of typhoons on cosmic-ray intensities, and during the Pacific War, greater emphasis was placed on such research, considered useful in understanding high-altitude atmospheric conditions. Part of the project would entail making very precise cosmic-ray counters, which would incorporate Geiger-Müller counters.

Discussions concerning an atomic bomb started as early as April 1940. In the early 1940s, Japanese physicists explored the use of nuclear fission during the war as an explosive or energy source. They studied the theory of chain reactions and conducted experiments to determine the neutron capture cross-section of uranium 235. Attempts were also made to construct apparatus for the separation of uranium isotopes by thermal diffusion and by the centrifugal process. The physicists came to the conclusion that it would be impossible to produce the atomic bomb during the war.[84] They suspected that it would be similar to a very compact and highly overcritical reactor, which might be built with uranium enriched in the isotope uranium 235.[85]

The project underwent four stages of development: (1) inquiries by the military, from 1940 to 1942; (2) a feasibility study by scientists from July 1942 to March 1943; (3) the Army funded "Ni-Project" carried out in Tokyo from late 1942 to April 1945; and (4) the "F-Project," which was commenced with Naval backing in Kyoto from mid-1943.[86]

Between 1940 and 1942, the military expressed great interest in the possibility of constructing an atomic weapon. Prime Minister Hideki Tōjō requested the chief of the Army Air Technical Laboratories (Rikugun Kōkū Gijutsu Kenkyūjo), Takeo Yasuda, to investigate the matter. He, in turn, passed the matter to Lt. Col. Tatsusaburō Suzuki, who subsequently produced a lengthy report in collaboration with Ryōkichi Sagane. Sagane had worked with Lawrence at Berkeley. Suzuki, who was attached to the Army Laboratories and also a member of Nishina's lab, had commenced studying physics under Shōji Nishikawa and Sagane at Tokyo Imperial University in April 1937. After speaking with Sagane and performing calculations, Suzuki submitted a 20 page report in October 1940, which concluded that an atomic bomb would be possible and that sufficient uranium might be found in Japan.[87]

The Suzuki report was not confidential and was distributed to various army and navy institutions, large companies such as Mitsubishi and Sumitomo, and university physics departments. Yasuda, keen to understand the principles of the bomb, invited Nishina and his students to give lectures on atomic physics to young soldiers. In April 1941, Yasuda formally requested Masatoshi Ōkōchi, Riken director, to conduct research into the construction of an atomic bomb.[88] Coincidentally, in that month's issue of the magazine "Chisei" ("Intelligence"), Yoshio Nishina called for scientists and technicians to help each other, to help defend the nation. The Navy, meanwhile, approached Bunsaku Arakatsu at Kyoto University to conduct their project.[89]

Nishina's article was certainly prompted by the war but what it reveals about the extent of his role in and contribution to the war effort is difficult to judge. Rhetoric such as Nishina's call to scientific arms may well have been a prerequisite for obtaining research funding at the time. Feigned patriotism or not, the reality was that wartime science enabled work on the cyclotron to continue.

Inquiries into the atomic bomb by the Navy were initiated separately through Captain Yōji Itō[90] of the Naval Technical Research Institute (Kaigun Gijutsu Kenkyūjo). Advice was obtained from Sagane and Professor Jūichi Hini and a Committee on Research in the Application of Nuclear Physics (Kakubutsuri Ōyō Kenkyū Iinkai) was, subsequently, established by the Navy and headed by Nishina. Over 10 meetings were held by the group of leading scientists from July 1942 to March 1943. The committee came to the conclusion that neither Japan nor the United States would be able to complete an atomic bomb during the war. Consequently, the Naval Technical Research Institute decided to concentrate its efforts in other areas such as radar research.[91]

The Army, which had provided funds for nuclear fission research at Riken from December 1940, continued to do so. It remains unclear what form the studies prior to 1943 took. Several scientists did make calculations and prepare data. Measurements were made of the fission cross-section of uranium. However, it was only from December 22, 1942 that a specific scientist, Masa Takeuchi was assigned to work on the problem. Takeuchi was a specialist in cosmic rays and hardly a nuclear physicist.

The research at Riken was dubbed the Ni-Project ("Ni-gō Kenkyū") after Nishina, its head. A list of personnel gives the names of 32 persons, although most contributed little to the project and some of those who did are not mentioned. Work was overall entrusted to the two young scientists Takeuchi and Kunihiko Kigoshi. Takeuchi designed and constructed an isotope separator while Kigoshi devoted himself to producing uranium hexafluoride. The product was to be introduced into the separator and be transformed into a gaseous state. It was hoped that upon heating, a quantity of uranium 235 would be surrendered. The investigations into isotope separation continued, without much success, until April 1945 when an air raid destroyed the building housing the gaseous thermal diffusion apparatus.[92]

A division of the Navy in the form of the Fleet Administration Center had, in the meantime, turned its attention to the problem. It provided funding for the Kyoto-based "F-gō Kenkyū" from May 1943, although participants' accounts of the date of commencement of research vary from late 1942 to early 1945. The F-Project was headed by Professor Bunsaku Arakatsu of Kyoto University; the letter "F" standing for "fission." The Kyoto group opted for an ultracentrifuge method of separating the uranium 235 isotope. A list of those participating in the project includes the names of Hideki Yukawa, Minoru Kobayashi, and Shōichi Sakata. All three are mentioned as working on theoretical research. Yukawa of Kyoto University was supposedly responsible for nuclear theory. Kobayashi, also of Kyoto, worked on uranium fission theory, and Sakata of Nagoya University investigated the theory of neutrons.[93] Although the project, on paper, received assistance from 19 scientists, there were only 5 principal contributors in addition to Arakatsu.[94] These were Kobayashi, who worked on theoretical problems, Kiichi Kimura and Sakae Shimizu, who were responsible for the centrifuge design, and Shinji Sasaki, who concentrated on chemical problems. Takuzō Okada studied the production of metallic uranium and was successful in obtaining a stable sample. The Kyoto project also resulted in a number of theoretical papers.

The efforts in Tokyo and Kyoto were hampered by a lack of required materials and facilities. In the latter part of 1944, there were discussions aimed at improving liaison between Army and Navy research groups, but the war ended before any major changes could take place.[95] Disorganization and red tape at the government level was also part of the problem, but the lack of sufficient supplies of uranium and other substances was a more major dilemma. A search for uranium was conducted by the Japanese Army throughout the new Asian empire, but to no great avail. Considerable sums were provided for research. Up to two million yen (around US$ 500,000)[96] was supposedly allotted by the Army to the Ni-project; how much was disbursed is unclear. One account suggests that the Army provided Nishina with 500,000 yen in 1942 and 700,000 yen for 1943–1945.[97] There are records indicating that 600,000 yen was provided by the Japanese Navy for the F-project.[98] Regardless of supposed funding, Takeuchi and Kigoshi were reduced to personally scrounging for basic materials for experiments. The tendency of the Army and Navy to make grants-in-aid to individual scientists rather than to enter into contracts with universities, as in the United States, fragmented the limited funds available.[99] Compared to the two billion U.S. dollar cost of the Manhattan Project, the Japanese project was very small indeed.[100]

Much valuable time was also wasted by Takeuchi and Kigoshi in attempting problems outside their scope of experience and knowledge. Expert advice was available from colleagues at Riken, but, in general, was reluctantly given. The Ni-project was characterized by overall ambivalence and lack of interest on the part of physicists.[101] At the time of Pearl Harbor, over 100 physicists were attached to the Nishina laboratory. Personnel much more qualified than Takeuchi and Kigoshi were available; yet, these people were not assigned to major roles in the project. They instead preferred to devote themselves to their own research.

It has been suggested that an atomic bomb project was also located in Nagoya. Due to bombing, it is said to have been moved to Konan, Korea, where a bomb was completed and tested three days before the end of the war. Apart from an American journalist's account of an interview with a Japanese officer who was an eyewitness, no other real evidence exists and little credence can be given to it.[102]

In addition to the atomic bomb, it was thought that physicists could contribute to the construction of a "death ray" by extending their work on radar. The plan was to construct a large-scale magnetron with an output of several thousand kilowatts. The output would, in turn, be concentrated via a reflecting mirror and aimed at targets such as planes. Microwaves emitted from the powerful magnetron would be transmitted and supposedly absorbed into the fat and bone of a human being, creating heat and hence destroying the tissue.[103]

But in radar development, like atomic bomb development, sectionalism and poor communication between the Army, Navy, and Ministry of Munitions resulted in wasteful duplication of research.[104] An extreme example of this can be seen in the division of Nippon Musen's manufacturing plant into two parts, one for the Army and the other for the Navy. After Japan's surrender, the U.S. Air Technical Intelligence Group stated in its report that

> Very severe criticism must be leveled at those Japanese military leaders who so long insisted that army and navy research, development, production and operation must be kept entirely separated. The number of scientists in Japan skilled to undertake radar research is much more limited than in the U.S.—it was inadequate to begin with. It was then the height of folly to insist on reducing their effectiveness by nearly one-half by requiring all projects, oftimes parallel, to be studied secretly within each of the two services.[105]

IV   The End of One Battle

In January 1945, Nishina won the prestigious Asahi Prize. In April, his laboratory was destroyed during an air raid.[106] The Department of Physics at Nagoya University dispersed and Sakata's theory group moved to Fushimi in Nagano prefecture.[107] In June, the U.S. secretary of war, Henry Stimson, invited a scientific panel consisting of Ernest Lawrence, Arthur Compton, Enrico Fermi, and Robert Oppenheimer to plan a demonstration of the atomic bomb, which might convince the Japanese to surrender. Unfortunately, they felt that "we can propose no technical demonstration likely to bring an end to the war; we see no acceptable alternative to direct military use."[108] On August 6, a uranium 235 atomic bomb fell on Hiroshima.

The young Home Ministry bureaucrat cum wartime naval paymaster officer, Yasuhiro Nakasone, was stationed at Takamatsu in Kagawa prefecture, on the island of Shikoku, at the time. Nakasone had graduated from Tokyo Imperial University in 1941, majoring in political science.[109] He was able to see the atomic cloud across the water of the Seto inland sea.

> I saw the mushroom cloud of the atomic bomb. That image will never fade from my memory. That lit a fire within me to develop atomic energy.[110]

In awe of the power that science could unleash, Nakasone resolved to involve himself, in the future, in political activities aimed at the promotion of science.[111]

Upon hearing the news of the devastation of Hiroshima, Nishina and Lt. Col. Tatsusaburō Suzuki traveled by a DC-3 plane to the site to confirm that it had indeed been an atomic bomb. They arrived in Hiroshima on August 8. Nishina sent various samples of debris back to Riken by an Army plane. On August 10, Motoharu Kimura measured the radioactivity of the samples that were received by Riken and confirmed that an atomic bomb had indeed been dropped on Hiroshima. Nishina's and Suzuki's trip was extended to Nagasaki on August 14 after it met with a similar fate, when a stronger bomb made of plutonium fell on August 9. Nishina proceeded to Osaka on August 15 where he heard the imperial proclamation of surrender, before returning to Tokyo.[112] Exposure to radiation would cost Nishina dearly. In late 1945, Nishina became bed-ridden, vomiting blood, experiencing stomach pains and internal bleeding. At the end of November 1945, the two Riken cyclotrons (and one each from Kyoto University and Osaka University) were confiscated and destroyed by the Occupation Forces. This was a great loss. Japan had more cyclotrons than any other nation outside of the United States and with their destruction, that lead in experimental physics was erased.

In February 1946, the awards continued with Nishina receiving the first Order of Cultural Merit to be given after the war, but this was little consolation for the destruction of his much-loved cyclotrons. Nishina was all too aware that he would have to go to battle once again if the Japanese were not to lose Riken as well as the war.[113]

The extent of the involvement of Japanese physicists in military research is extremely difficult to gauge. Yasutaka Tanikawa mentions that Yukawa "was supposed to be working for the military but he did not."[114] Postwar reflections on wartime research tend toward the view that research was conducted under pressure and done in order to prevent young scientists from being sent to war. There were, thus, both nationalistic and personal imperatives to ensure participation of scientists at some level. Military research meant self-preservation for younger scientists faced with the option of going to war. Nishina and other senior scientists engaged in patriotic rhetoric and activities at a higher organizational level, possibly in an attempt to save younger colleagues from such a fate.[115] However, these imperatives never went beyond engendering a token and superficial enthusiasm. The F-project was still at the theoretical stage at the close of the war. Research was, furthermore, conducted in the belief that a bomb was unlikely to be completed, let alone used against the enemy.

Accordingly, few Japanese scientists were purged on the grounds of cooperation in the war. Nishina was not one of them. Most of those prosecuted were involved in technological support of the military.

They did, however, include the astronomer Toshima Araki, professor in the science faculty at Kyoto University.[116] A conflict of values arising from political and ideological beliefs led to less-than-active participation. Mituo Taketani carried out calculations for the Ni-project while detained by the Thought Police. However, as there was little likelihood of producing a bomb, physicists were able to accommodate the demands of the policy-makers while maintaining their own values. It appears that the scientists treated the request to build an atomic weapon more as an assigned scholarly project.[117] They saw it as a means by which related research could also be funded and conducted under the banner of building a bomb. Research was thus confined to nothing much more than an academic exercise.

There was very little contact between scientists in Germany and Japan in the case of both bomb and radar research.[118] But although the Japanese projects would undoubtedly have benefited from German opinion, Japanese intellectual resources were not even used to the full. Postwar reflections by participants have a tendency to bathe accounts of wartime research in a sympathetic light. However, the reality was a small-scale effort lacking in manpower, coordination, and materials.[119] Even with the restrictions of wartime Japan, more could have been achieved, but was not.

The mobilization of intellectual resources involved an abrupt adjustment on the part of scientists. The extent of this adjustment is difficult to gauge, as much of what was carried out under the name of "wartime research" was related to pure science. The ambivalence of some physicists toward military work can be traced to academic values and feelings of social responsibility. This, however, never appears to have gone beyond passive resistance. Ironically, the war years are remembered as the "good old days," when funds and research were the least restricted.[120] In a survey conducted by the Science Council in 1951, a number of scientists replied that the Pacific War period was the time of their greatest freedom in research.[121] Military research, or for that matter applied research, hardly constituted a "sellout" by pure scientists.

Despite Japan's defeat, the war was "useful" for science and engineering. Compared to a decade earlier, science and engineering graduates tripled during the Pacific War. Some graduates were absorbed in the expansion of domestic research facilities in both basic and applied research. Riken was a major example of this. The military effort also required that manufactured goods be of uniform quality, encouraging Japanese industry to adopt standards for mass production.[122]

Nishina, who had never occupied the ivory towers of academia, was well located to take advantage of the new opportunities in funding that became available. Justification for his research co-opted official discourses regarding the need for a national defense state. When Nishina pressed the Army liaison officer Major General Nobuuji for assistance in completing the 60-inch cyclotron in 1943, he used the following words:

> The 250-ton, 1.5 metre accelerator accelerator is ready for operation except for certain components which are unavailable as they are being used in the construction of munitions. If this accelerator is completed we believe we can accomplish a great deal. At this moment the U.S. plans to construct an accelerator ten times as great but we are unsure as to whether they can accomplish this.[123]

Many "public" scholars such as Nishina found it useful to support the war effort, yet, at the same time, express passive resistance by continuing research that had little by way of military applications.[124] The public interest in Japan, which physicists such as Nishina and Nagaoka claimed to serve, was the same national interest that required a strong military. It is no coincidence that the rhetoric overlapped. Nishina's ability to negotiate with those in power would see him in good stead after the war. Not all scientists had such talents. After Japan's surrender, President Harry Truman was told by Karl T. Compton, who had been part of the U.S. Scientific Intelligence Survey to Japan, that there had been a lack of liaison between military organizations and university scientists because "the technical staffs of the military organizations were afraid to discuss their technical problems with the more competent university scientists lest their own incompetence be disclosed."[125]

CONCLUSION

This chapter has outlined the career of Yoshio Nishina till the end of World War II. As Yukawa's poem so aptly expressed, the path of science was intertwined with the notion of service to the nation. The mobilization of science and the role of Nishina and members of his laboratory show how physicists were able to stake a claim to military research funding, achieving modest success in the exploration of nuclear fission. Nishina's enthusiasm for the cyclotron, and the rhetoric that accompanied it, show how fundamental studies of nature could be manipulated for the purposes of national power and prestige.[126] He turned to the military for research support and was able to skillfully balance wartime agendas with academic interests.[127]

Even though he did not come close to completing a bomb, an understanding of the atom brought with it political influence. Although the Pacific War meant a temporary stop to the internationalism of science, the development of Yukawa's meson theory continued. Also, paradoxically, the war with the Americans enabled Nishina to continue the Americanization of Japanese physics via his cyclotron building.

In addition to selling the ability of physicists to produce "weapons of war," Nishina was able to secure a special place for physicists in society by rapidly mastering the organizational culture of big science, which included the administration of large teams of researchers, a large budget, and liaison with government agencies, symptomatic of the needs created by Berkeley big science. Nishina and Nagaoka influenced policy-making through a number of channels, including membership of committees, direct contact with military representatives, and good public relations. Ernest Lawrence's argument that physics should be funded because of the public good was often used by Nishina and those who followed him. Self-interest (in terms of doing physics), however, seems to have been foremost in Nishina's mind up until the end of the war. It is no coincidence that after the war, Nishina would be called in to save Riken from financial disaster and possible dissolution by the Occupation forces. What is more, Nishina would be consulted in the reorganization of science and technology by the Occupation authorities themselves. His international network of scientific contacts would also ensure that his voice would be heard far beyond Japan's borders.

In sum, this suggests that physicists were well placed to influence science policy and Japan's reconstruction at the end of World War II, not unlike their colleagues in the United States. We saw earlier how physicists were shaped by culture and institutions. This chapter showed how, in wartime Japan, there was a tension between loyalty to the nation and a commitment to the internationalism of science. Physicists had multiple identities. In the following chapters, we will see how internationalism came to the fore when Nishina and other physicists sought to shape the postwar world, in the context of a "democracy" brought to them courtesy of the Americans.

CHAPTER 3

THE IMPACT OF THE ALLIED OCCUPATION: NISHINA AND NAKASONE

Jean-Jacques Salomon has written that throughout the world,

> at the end of the war, the demobilization of researchers, far from
> signalling the end of "mobilized" science as such, gave rise, on the
> contrary, to systematic efforts to take advantage of research activities in
> the context of "national and international objectives."[1]

Japan was no exception. Indeed, after Japan's defeat, there were immediate calls for the building of a new Japan, a "scientific Japan." In the August–September 1945 issue of the popular science magazine *Kagaku asahi*, Professor Kinjirō Okabe of Osaka University called for a new start in scientific research. He envisaged a Japan that could contribute greatly to a world culture ("sekai bunka") by producing the likes of scientists such as Hideyo Noguchi, Hantarō Nagaoka, Kōtarō Honda, and Hideki Yukawa.[2]

Both Hantarō Nagaoka and Yoshio Nishina played significant roles in the mobilization of science during the war, working with the military and government bodies. Nishina also played an important role in the "remobilization" of science and technology afterward as well. But the bulk of scientists had less access to policy-making. Scientific bodies in Great Britain, the United States, France, and Japan, all called for the establishment of central organizations for science policy,[3] which would assist in postwar reconstruction and provide less well-connected scientists with a means of ensuring access to government.

Many Japanese, including scientists, considered that weaknesses in science and technology, ultimately, had cost the Japanese their final victory. Japan had not been able to build an atomic bomb. A long list of contributing factors included inhibiting secrecy, interservice and interdepartmental rivalry, and institutional factionalism. These same criticisms had been made during the war. Afterward, they were used in the context of targeting science and technology toward the reconstruction of a "peaceful" Japan and revitalization of the Japanese economy. This time, the only difference was that the United States was calling the tune.[4] Americans influenced Japanese physics before the war but the Allied Occupation provided an extended opportunity for the Americanization of all facets of life, the prelude to a long-term alliance in which Japan has been likened to a U.S. protectorate.[5]

The first three sections of this chapter describe how the Occupation authorities sought to demobilize Japanese science, remobilize it to meet U.S. goals, and reorganize scientific bodies along more democratic lines. While Nishina features throughout these sections, the fourth focuses on his activities during this period.

I DEMOBILIZING THE JAPANESE

On August 8, 1945, Nishina accompanied a group from the Imperial Headquarters to Hiroshima to confirm the first bombing.[6] A Navy survey group, which included Bunsaku Arakatsu, followed, and on August 30, groups from Tokyo and Kyushu Universities arrived in Nagasaki. On September 5, a general investigatory group from Kyoto University arrived in Hiroshima, to be joined on September 16 by a group from the Faculty of Science. On that day, in Tokyo, the National Research Council (NRC) established the Special Committee for the Investigation of the Effects of the Atomic Bomb. This committee dispatched a survey group to Hiroshima. In October, representatives of the U.S. Army arrived. The Japanese scientists cooperated with the U.S. authorities and submitted the results of their research to the headquarters of the Supreme Commander for the Allied Powers (SCAP) for its approval for publication.[7]

The Japanese government accepted the Potsdam Declaration of July 26, 1945, which stated that it should "remove all obstacles to the revival and strengthening of democratic tendencies among the Japanese people." SCAP interpreted this accordingly and was not aware of any legal limitations on Occupation legislation, given that Japan had surrendered "unconditionally." For SCAP, there appeared to be no boundaries to its authority.[8] Two other Allied bodies did exist, the Far Eastern Commission in Washington (FEC) and the Allied Council for Japan in Tokyo, but these only had nominal responsibilities for formulating and reviewing policy-making.[9]

The Japanese were required, by the terms of the U.S. Initial Postsurrender Policy, to maintain all facilities, apparatus, and materials associated with its war effort, and to await directives from SCAP as to their disposal. It was, however, 20 days after the end of the war, on September 4, 1945, that the Agency of Science and Technology, one of the pillars of Japan's wartime mobilization, was dissolved, and its functions reallocated to the newly established Cabinet Investigation Bureau, the Science Education Bureau of the Ministry of Education, and the Patent Standards Bureau of the Ministry of Commerce and Industry. On August 26, the Ministry of Munitions and the Ministry of Agriculture and Commerce also disappeared, with the earlier Ministry of Commerce and Industry and the old Ministry of Agriculture and Forestry taking their place. This was the day before the Occupation Forces arrived in Japan.[10]

Back in the United States, "New Dealers"[11] such as Senator Harley Kilgore proposed that a National Science Foundation be established to support research.[12] The New Deal spirit is evident in Japan, too, in the attempts by the Economic and Scientific Section (ESS) of SCAP to set up a democratic science policy-making structure, which would both coordinate and control Japanese scientists. The ESS established a formal governmental organization to achieve this, but to its later dismay, the bureaucracy ran out of control, creating a widening gulf between "democratic" and government structures. Power remained in much the same bureaucracy that had organized prewar and wartime science and technology. As the official, "restricted" history makes clear, "there was a continuing SCAP conflict between the view of the inherent danger of a strong science and technology and the view of its necessity to a healthy economy and self-respecting nation."[13]

In September 1945, the U.S. Scientific Intelligence Survey group was sent to Japan to investigate wartime activities.[14] The U.S. Initial Postsurrender Policy prohibited any research and/or instruction that could be construed as contributing to a revival of Japan's war-making potential. Fear of a possible Japanese atomic bomb ensured that all research activities relating to atomic energy were prohibited, all such facilities were seized, and personnel interrogated. Facilities were to remain closed until SCAP could ascertain that activities were of a peaceful nature.[15] Instructions were given by SCAP to destroy or scrap "enemy equipment"—arms, war vessels, aircraft, and military installations. There were certain specific exemptions, including

(1) equipment considered "unique and new development" desirable for "examination, intelligence or research"; (2) equipment deemed useful for U.S. army or naval operations; and (3) equipment that was thought to be suitable for peacetime civilian use.[16] Research was continued in wave propagation at the Radiophysics Research Institute (October 10, 1945); on ocean observation (February 8, 1946); and on mapmaking (May 9, 1946). Some facilities and laboratory staff, such as those of the Department of Chemistry at Tokyo University, were also used.[17]

The "demilitarization" of Japanese science and technology entailed the temporary control of all military research facilities and aeronautical laboratories, some of which were converted to establishments for peaceful purposes. Attempts made by the Japanese Government to attach the Central Aeronautical Research Institute to the Railway Research Institute of the Ministry of Transport, for example, were eventually approved in October 1947, and aeronautical facilities were turned to civilian use. Similarly, approval was granted for the return of the Shimada laboratory of the Second Naval Technical Research Institute, for use by the Government Electro-technical Laboratory.[18]

Negotiations led to the return or transfer of equipment, such as meteorological instruments owned by the Army, to the Central Meteorological Institute.[19] When equipment used in military laboratories was suitable as reparations, such was not always available for civilian use. At Tokyo University, the Aeronautical Research Institute was able to be reorganized as an "Institute of Science and Technology," although its equipment was open to possible confiscation as reparations.[20]

As late as mid-1946, aircraft were considered "implements of war." The strict prohibitions on R&D and instruction were slightly relaxed when SCAP approval was obtained for the publication of wartime research in aeronautical science. The SCAP opinion was that:

> it is both dangerous and futile to attempt to keep any people in darkness about advances in scientific knowledge . . . it would seem possible to distinguish between knowledge and harmful application of knowledge, the latter of which can be prohibited without paralysing the former.[21]

Toward the end of 1945, after initial investigations had ascertained that there was no Japanese A-bomb, restrictions on Japanese nuclear physicists were relaxed, and orders to guard Japanese laboratories were lifted. But the prohibitions on research relating to atomic energy remained, and personnel with any knowledge of the topic were registered and kept under surveillance. Continued fear of Japan's atomic potential, especially given the high level of Japanese theoretical physics, led to the impounding of all uranium and thorium with zealous thoroughness.[22]

In a letter to General MacArthur dated October 15, 1945, Nishina requested permission to operate his cyclotron for studies in biology, medicine, chemistry, and metallurgy. Permission was, at first, granted, but later withdrawn. Nishina's request was the only case in which a research program was discontinued because of SCAP objections. Other Japanese scientists conspicuously avoided research that could be construed as violating Occupation guidelines. Japan's four cyclotrons were seized on November 20 and, four days later, dismantled and dumped into Tokyo Bay.[23] Two of these machines were at Riken, one at Osaka University[24] and the other at Kyoto University. *The New York Times* referred to these locations as "atomic plants."[25] The order for the destruction of the cyclotrons has been attributed to General Leslie R. Groves.[26]

The U.S. National Academy of Sciences and the Association of Oak Ridge Scientists protested strongly against such wanton vandalism. The U.S. Army dispatched Gerald Fox and Harry Kelly, physicists who had been attached to the Radiation Laboratory established by the U.S. Office of Scientific Research and Development (OSRD) at MIT. Their mission was to liaise with Japan's scientists. At the beginning of 1946, they were assigned to the Science and Technology Division of SCAP. Fox was to stay less than a year but Kelly remained for four years, during which time he became deputy chief of the division.[27] Kelly's job also included heading a Special Projects Unit in ESS, which censored material dealing with the atomic bomb.[28] Kelly and Nishina agreed that science should be mobilized for the benefit of Japan. It was no coincidence that the Science and Technology Division was placed in SCAP's ESS.

Kelly acted as a conduit for Japanese physicists, providing them with a link to the outside scientific world. Lawrence sent Nishina a number of journals through Kelly. As Nishina explained to Lawrence,

These journals are the only channel through which we can get in touch with the scientific world abroad [*sic*]. We were completely cut away from the [*sic*] international science since the summer of 1941 and now we are gradually getting familiar with it.[29]

Nishina, the consummate politician, attempted to use the destruction of the cyclotrons and his own international network to elicit sympathy and favorable treatment for Japanese scientists. A short account of the destruction of the machines he gave to Drs. Henshaw and Bruce of the Atomic Survey Group in late 1946 had the desired effect of showing the stupidity of SCAP's actions. Harry Kelly felt that Japan had little need of a cyclotron, and that Nishina was trying to be recognized as "a martyr of science."[30]

Nishina's wartime experience taught him that science could not be divorced from politics. On March 12, 1946, he put pen to paper and called for a union-type organization, which would represent the interests of scientists throughout Japan. Through political action, it was hoped that scientists could be assured of, at least, the barest minimum by which to survive. Such an organization could be used for other purposes as well:

Also, this association could help in the political training of scientists. The current political awareness of scientists is insufficient. Or it might be better to say that the majority of scientists are indifferent to politics. This constitutes one of the major reasons for the stagnation of science in this country. As democratic politics will be party politics which reflects the views of the people, political parties will hold the key to the rise and fall of science in the future. Furthermore, we need to give serious thought as to which party should take charge of science policy. For this reason, I would like each party to express its aspirations with regard to science policy.[31]

Nishina's essay was published in the first issue of the science magazine *Shizen* (*Nature*), which came out in May 1946. The change in language is rather remarkable, not unlike the words of a union organizer. What Nishina essentially called for was for scientists to translate their authority into political clout through representation and participation in the policy-making process. Nishina qualified his remarks later that year with a new sense of urgency. In the December-1946 issue of the same journal, he lamented the rise in crime, lack of food, and low production levels. Science and technology, he felt, had an important part to play in solving some of the problems facing Japan. This took priority over politics:

Of course, we should, as citizens, work towards the promotion of morals, character building, and the advancement of cultivation amongst people. This underlies everything, but as scientists and technologists it is our duty to make earnest efforts to revitalize industry. In order to do this as a nation, one central organization may be necessary. However, if such an organization fails, we will be left with nothing, so what needs to be accomplished first is realizing the goal. We must make new things with new methods by using our respective specialist knowledge and applying it to the creation and development of manufacturing technology. If the revitalization of industry follows a rising curve as a result of such measures, people will have a more comfortable life and be more courteous to one another, production will increase and this will stimulate progress in science and technology.[32]

The "scientific method" was viewed as a means by which to identify and solve social problems, and it was through a science policy which emphasized the importance of applied over basic research, and the importance of commercialization that Japan could be reconstructed. Physicists were being positioned to play the role of "experts" and charged with the task of helping to rebuild Japan. Amidst Japan's defeat, it is not surprising that these talented young men would provide leadership as public figures. But the government had to be made aware that they needed such expertise. A representative organ for scientists and scholars was viewed as one way in which this could be done. It would also serve as a means of monitoring their activities.

II   REMOBILIZING THE JAPANESE: SURVEILLANCE AND CONTROL

In May 1946, in order to maintain control over Japan's scientific activities, Brigadier John O'Brien, the Australian chief of the Scientific and Technical Division of ESS proposed that SCAP employ ten scientists to report on developments, and to relay instructions of the Division. He also suggested a regular research forum for Japanese scientists and scientific representatives of the Allied Powers to discuss the policies and problems of research institutions. Such a forum could "make no decisions" but it could make recommendations, and the later establishment of scientist liaison committees were a result. This strategy relied on Japanese goodwill and was

> based entirely on Japan being "good boys." How far can we go? To what extent are we to trust the Japs.[33]

It was reported that there was excellent cooperation on all sides.

The surveillance of Japanese R&D was facilitated by a directive, in September 1945, for the submission of monthly research reports. This was found impracticable and in May 1946, the directive was revised to six-monthly reports and in May 1949, to annual reports. Apart from prohibitions on research of an overtly military nature, the only restriction that was placed was on atomic energy research. The national

Committee for Nuclear Research, chaired by Nishina, was set up with the principal aim of preparing these reports.[34] It later evolved into the Science Council's Special Committee for Nuclear Research.

As a further part of its information-gathering activities, SCAP requested the Japanese Government, in March 1947, to provide biographical data on all scientific and technical personnel capable of, or active in, R&D activities. The usefulness of Japanese scientists was becoming more apparent.[35] Later that year, a U.S. Scientific Advisory Group was sent to Japan by the U.S. National Academy of Sciences, after a SCAP invitation and in connection with the moves by SCAP to establish a democratic and effective science organization in Japan. The scientific mission consisted of Roger Adams (University of Illinois), Merrill K. Bennett (Stanford University), William D. Coolidge (General Electric Company), William V. Houston (Rice Institute), William J. Robbins (New York Botanical Garden), and Royal W. Sorensen (California Institute of Technology). The group submitted a report entitled "Reorganization of Science and Technology in Japan" dated August 28, 1947. The group's recommendations, it seems, were used only for reference.[36]

The political climate had changed considerably by late 1948, when a second scientific mission was sent to Japan. This group, consisting of Roger Adams, Detlev W. Bronk (Johns Hopkins University and National Research Council), Zay Jeffries (General Electric Company), I.I. Rabi (Columbia University), and Elvin C. Stakman (University of Minnesota), arrived on November 28, 1948 and stayed for three weeks. They made six formal recommendations, requesting (1) the establishment of national fellowships; (2) Japanese funds to send advanced Japanese students overseas to study; (3) the international exchange of senior scientists; (4) the use of part of Japan's export earnings for improvement of scientific laboratories; (5) the encouragement of practical applications of science; and (6)

> the development of a few outstanding centers of scientific research and education as standards for the ultimate improvement of scientific research throughout the country rather than the dissipation of limited funds among many institutions.[37]

I.I. Rabi, Nobel- Prize physicist for 1944 and professor at Columbia, had worked with Nishina in Germany before the war. In addition to being chairman of the Department of Physics at Columbia, he was a member of the General Advisory Committee of the Atomic Energy Commission and trustee of the Brookhaven National Laboratory (figure 3.1).[38]

Figure 3.1 Yoshio Nishina and I.I. Rabi at a welcome party for the U.S. Scientific Advisory Group, Tokyo, December 1948.

Source: Nishina Memorial Foundation. Courtesy, AIP Emilio Segrè Visual Archives.

During the years 1942–1945, Rabi was associate director of the MIT Radiation Laboratory, the location of radar development and where Kelly and Fox had been working.[39] Like a number of American physicists, Rabi had recently become acquainted with Japanese advances in theoretical physics. He held Japanese physicists in high esteem,[40] but was taken aback by the poor living and research conditions. On December 8, 1948, just seven years after Pearl Harbor, Rabi recommended to Kelly that

> What we have seen of Japanese research laboratories and Japanese scientists indicates to me that it would be very efficient from the point of view of the production of scientific results which would be useful to the United States to increase the aid to Japanese science. . . . Relatively small sums would put much of this expensive equipment and these highly trained individuals to work in an efficient fashion to produce scientific results of value. For example, in Professor Sagane's researches at Tokyo University, the addition of $2000 or $3000 to the budget would enable them to get their equipment into effective operation very quickly and they would produce results of value. The same is true of other places.[41]

Although the memorandum sounds a highly philanthropic one, U.S. interests were also in mind. In another memorandum of the same day, Rabi recommended the limited "importation" of Japanese scientific personnel to the United States for two or three years. Rabi rationalized this in the following way:

> Japan in a number of fields, has scientists who are now ineffective because of the lack of equipment. In the United States, on the other hand, we have ample equipment and support but suffer from the shortage of scientific personnel. There are many cases where Japanese scientists would be very useful in researches in the United States, which are supported from military funds given to universities. In many cases these funds are inefficiently used because of the shortage of highly trained personnel in the United States.[42]

An abundance of Japanese talent had little in the way of research facilities. Inviting select Japanese scientists to the United States would benefit both sides. This was consistent with the thinking of General William F. Marquat, chief of the ESS, and Kelly:

> It was decided that if Japanese scientists were to return to the United States, they should return as scholars and sponsored by some civilian agency who would be responsible for security measures, rather than to import them into the United States as was done in Germany. The adverse publicity, even by American scientists themselves, due to importation of the German scientists has proved our method of approach to be more feasible.[43]

Kelly mentioned Yukawa's departure on September 2, 1948 to accept an appointment in the Institute of Advanced Study, Princeton. He hoped that in the future, Marquat might also permit other Japanese to follow:

> With your approval, the same system will be followed in the future, and every encouragement will be given to the travel of outstanding scientists of Japan to the United States rather than to unfriendly nations.[44]

On December 10, 1948, Rabi's proposal took the form of an official, "top secret" memorandum to Marquat. Entitled "The Use of Japanese Research Facilities as an Advanced Base in the Event of Acute Emergency in the Far East," it related the potential usefulness of Japanese scientists, particularly in time of war. Rabi cited the advantages of having a British base close to the actual scene of operations. He recounted the Telecommunications Research Laboratory at Malvern in England. By converting a laboratory to an American-manned laboratory, American forces had access to new equipment at an early stage of development and production.[45] One or more such laboratories in Japan, in fields such as physics and telecommunications, would provide the United States with a similar strategic base. Because of the distance and logistic difficulties, the Japanese were to immediately establish a laboratory with U.S. assistance. He concluded with the statement that "Such laboratories would also have an important favorable effect on the Japanese economy in a direction favorable to American interests."[46]

In the principal fields of science, 18 Japanese laboratories were suggested as possibilities. One of the main criteria was the applicability of research to aiding Japan's economic recovery. Of these, 8 were considered

> worthy of receiving active assistance from the United States through importation of particular types of specialized scientific equipment and other means in order that they might better qualify to serve as base laboratories to be used by the United States in the event of an emergency. Preliminary investigations indicate that only a small amount of equipment would be required for each laboratory in order to increase its effectiveness to the point desired.[47]

The 8 laboratories singled out for special assistance included Nishina's Scientific Research Institute. Rabi assessed Nishina's laboratory as being one of the better research organizations, with 3 locations in Tokyo and a research staff of 192 university graduates and 223 technical assistants working in the applied areas of physics, chemistry, mechanics, and antibiotics.[48] SCAP policy had, to that time, stressed research activities of a nonmilitary nature.

The General Staff Section G-4 informed Military Intelligence G-2 that they were of the opinion that the proposal would benefit the United States but they feared that

> Unless surrounded with the utmost secrecy and its purpose adequately concealed, the project might be used by powers unfriendly to the U.S. to prove that SCAP (and D/A) were furthering military research in Japan or were even allowing Japan to conduct its own military research in preparation for war.[49]

The top secret document was marked "Colonel Rash" [sic]. It is possible that this actually refers to U.S. Army Colonel Boris T. Pash who is known to have served in Occupied Japan under General Douglas MacArthur.[50] Pash had investigated communist activities among staff members of Ernest Lawrence's Berkeley laboratory,[51] and served as chief counterintelligence officer on the Manhattan Project, the U.S. atomic bomb program headed by General Leslie R. Groves and J. Robert Oppenheimer who was director of the Project's Los Alamos Laboratories. Groves appointed Pash as head of the Alsos mission sent to Europe in 1944–1945. Alsos was Greek for "grove." Pash and the physicist Samuel Goudsmit, who was scientific head of the mission, investigated German efforts to produce an atomic bomb, as well as probing into other military-related research and development.[52] Several years later, in 1954, Pash's testimony against Oppenheimer would see Oppenheimer classified as a security risk.

On March 19, 1949, Marquat wrote to the Department of Army informing them of his approval of the "Base Laboratories in Japan" program. What became of the program is not known.[53] After discussions with Kelly regarding the future activities of the Scientific and Technical Division, Marquat recommended that the Japanese be encouraged to establish a number of laboratories, especially in physics and communications, which would be staffed by Japanese scientists of high caliber. Such laboratories could be used by the United States in times of emergency, as was the case with British facilities that were used during World War II.

The six recommendations of the U.S. Scientific Mission, the official SCAP history states, were implemented by the time of the end of the Occupation; but "there was no conscious effort to build up a few chosen laboratories as exemplary research centers."[54] It remains unclear why this was so. Ideas of an overseas scientific base were not, of course, unique to Rabi. In writing of his trip to Japan in late 1945, the MIT physicist Karl T. Compton described the "development of technique [*sic*] for making scientists useful in the active theaters," which he traced back to the establishment by the U.S. OSRD of a London Liaison Office for British and American scientists. This office provided support in Europe for scientists and engineers who had developed weapons in the United States. Compton cites the British Branch of the Radiation Laboratory for Radar (operated by MIT) and the American–British Laboratory-15 for Radar and Radio Countermeasures (Harvard University) as being particularly important. He spoke from experience, as he was both director of the Pacific Branch of OSRD and president of MIT.[55] In 1951, L.N. Ridenour wrote that science was "the shield of the free world." For Compton and Rabi, Japanese science had a part to play in mobilizing for the defense of that freedom as well.[56]

III REORGANIZING SCIENCE

The initial nurturing of "democratic tendencies" amongst the Japanese was seen as ensuring that Japan would not reemerge as an aggressive power. It was hoped that democratic control of science and technology could be exercised through governmental bodies and democratic scientist organizations.[57] The Japanese were to be encouraged to devise a plan for reorganization themselves, subject to review by SCAP. Interested staff sections of SCAP, such as Public Health and Welfare (PH&W), Civil Information and Education (CI&E), ESS, and Natural Resources Section (NRS), would all act as "godfather" to the proposed organization.[58]

In March 1946, following representations to the Minister of Education by the NRC,[59] a body to promote scientific reorganization was formed. This consisted of ten members drawn from the three main scientific organs: the Imperial Academy, the NRC and the Japan Society for the Promotion of Science (JSPS). It soon proposed to dissolve the NRC, increase Academy membership and activities, and make JSPS more compatible with a revitalized Academy. Many scientists were unhappy with the plan, but until a more appropriate organization could be established, it was decided that the "big three" should continue their work.[60]

The Government Section of GHQ was dubious about the three main scientific bodies, not least because of their involvement in the war effort. On June 5, 1946, Japanese representatives of scientific organizations gathered at the Central Liaison Office of the GHQ to hear an address by Brig. John O'Brien. Preliminary discussions had indicated that the establishment of a closer liaison between certain sections of SCAP and Japanese science might prove productive and Harry Kelly had been appointed to assist the Japanese in their deliberations. By organizing a liaison committee, O'Brien felt that it might "eventually grow into a democratic organization of Japanese science and scientists having objects that are major contributions to the proper rebuilding of Japan."[61]

It was, thus, between April–June 1946 that the Japanese Association for Science Liaison (JASL) (Kagaku Shōgai Renraku Iinkai) was formed, primarily as an advisory group to service the Occupation. On June 5–6, 1946, Kelly and O'Brien met with representatives from the major universities, Riken and the Ministry of Education. A "Japanese Association for Science Liaison" was decided upon, with a central office located in the Ministry and branch offices in each of the seven imperial universities. Later, steps were taken to widen membership to include representatives from medicine and engineering. On October 28, Sagane suggested that a similar group specifically for liaison with engineers be established, and by December, the Engineering Liaison Group was in the process of being formed.[62] The JASL consisted of a network of regional branches at imperial universities in Hokkaido, Tohoku, Tokyo, Nagoya, Kyoto, Osaka, and Kyushu. The 40 members included Yoshio Nishina. The Association continued its activities right up to the establishment of the Science Council of Japan.[63] Some members of the Association viewed themselves as being

> just like a member of a Travel Bureau in the field of physics, chemistry[,] biology and the fundamental researches of engineering, by solving quickly the problems which are thought to be hopefull [*sic*], chiefly among the problems between G.H.Q and Japanese scientists.[64]

The JASL saw one of its roles as impeding plans for scientific reorganization put forward by less democratic, nationalistic groups, which ran counter to the efforts of SCAP. In June 1946, an organization called the "Scientific and Technical Policy Comrades Association" ("Kagaku Gijutsu Seisaku Dōshikai") was formed under the leadership of Hidetsugu Yagi. Yagi (of Yagi antenna fame) headed the Agency of Science and Technology during the war and was president of Osaka Imperial University until being purged in 1946. The Association was instrumental in setting up the Dietmen's Science Club, a group of some 70 members who were interested in the promotion of science for political purposes. On November 19, 1946, Cabinet approved the formation of an official committee to carry out the nationalist policy being advocated by the Comrades' Association. SCAP objected and the Association came to a (perhaps temporary) halt in December 1946. As the official SCAP history perceptively comments:

> To those who were resisting basic changes in the structure of science organization, there were inherent in the plan (to democratically reorganize science) attractive time-consuming complications: it was conceivable that the Occupation would end or be reduced in influence well before the completion or firm establishment of a drastically changed science organization; it is possible also that the entrenched leaders felt that they could wear the cloak of democracy and still not lose the position of dominance which was theirs traditionally. Events showed that they tried.[65]

Although it had been intended that it would encourage use "of science and scientists more effectively to meet with the national needs,"[66] it became apparent that the JASL included little representation from the applied sciences.[67] A high-level committee with control of all technical research and its application to industry, such as Yagi may have had in mind, might have been more effective, but GHQ was anxious that such a concentration of control over funds in one board might lead to the emergence of a new technical general staff. As a SCAP report commented, this "illustrated the ever-conflicting requirements for the elimination of war potential and the needs for the rehabilitation of Japan."[68]

Nevertheless, as Major Roche, assistant executive officer of the NRS commented at a meeting held by the Scientific and Technical Division on April 2, 1947, "it would be easier to keep surveillance if there was one centralized body of scientists."[69] The physicist Satoshi Watanabe wrote in his "Suggestions for Rehabilitation of Scientific Activities in Japan" in July 1947, that only a combined effort on the part of industry and science could lead to a national economic recovery. A committee consisting only of scientists would be powerless. While he felt that "science for the people" was important, he also felt that it would have to be supported by the people as well.[70] The "Science Council," which finally evolved, represented the compromise that was made: a centralized body with limited power and effectiveness.

A preparatory committee (Sewaninkai), consisting of 44 mainly Tokyo-based academics, was established in January 1947 to be responsible for electing a representative Renewal Committee for Science Organization (Gakujutsu Taisei Sasshin Iinkai). Following deliberations of this committee, the Science Council of Japan was established. The Renewal Committee was to be made up of 108 members, with 15 members each coming from the following 7 areas: basic natural science, engineering, medicine, agriculture, law, literature, and economics. The remaining 3 would form a special "composite" category representing broad interests covering 3 or more of these fields.[71] Despite a proposal made by the Preparatory Committee, the Renewal Committee chose not to include additional members from the Imperial Academy, the NRC, the JSPS, government offices, and the Diet. However, in order to obtain government funding, it was thought best for the Committee to be established within the NRC.[72]

Progressive groups were active during the election in 1947. Eleven communist sympathizers were elected, two of who belonged to the composite category. These two members represented what SCAP called "communist-front" organizations—the Democratic Scientists' Association (Minshushugi Kagakusha Kyōkai or Minka) and the Democratic Technologists' Association (Minshushugi Gijutsusha Kyōkai). The Renewal Committee continued its activities until March 1948.[73]

GHQ had little opportunity to influence the course of the Renewal Committee until its final months, when pressure was mounted on the relationship between the proposed Science Council and the government. The Tetsu Katayama Cabinet fell on February 10, 1948 and was replaced by one led by Hitoshi Ashida a month later. Amidst this political change, major decisions were made. To meet anxieties that government would allow a left wing takeover of the Science Council,[74] it was proposed that the Council should be independent. After consultation with GHQ, it was decided that in the act creating the Science Council, the phrase "must consult" would be changed to "be able to consult," as the former would unduly "constrain" the government. The government could seek the advice of the Science Council of Japan, but it was not bound to do so. (In later years, the government increasingly turned to its "own" advisory bodies for scientific opinion.) Furthermore, the Japan Academy (known previously as the Imperial Academy) was to be contained within the Council, with the Council having the authority to select its members.[75] This decision overruled protests from Hantarō Nagaoka, the president of the Academy,[76] and his successor, the senior law specialist Saburō Yamada.[77] To Nagaoka, there was the indignation of having to see a younger generation of physicists come to the fore. The institution that he led and the elitist values that it espoused were transformed.

In the United States, too, there were similar concerns that the establishment of the National Science Foundation might give civilian scientists too much control over policy. President Harry S. Truman vetoed a bill in 1947, stating that:

> this bill contains provisions which represent such a marked departure
> from sound principles for the administration of public affairs that I can-
> not give it my approval. It would, in effect vest the determination of
> vital national policies, the expenditure of large public funds, and the
> administration of important governmental functions in a group of indi-
> viduals who would be essentially private citizens. The proposed
> National Science Foundation would be divorced from control by the
> people to an extent that implies a distinct lack of faith in democratic
> processes.[78]

The need for "democracy" was thus used to limit the autonomy of
scientists, as in Japan.

The recommendations of the Renewal Committee were passed by
the National Diet on July 5, 1948. The Science Council of Japan
would henceforth consist of 210 elected members who, in turn,
would belong to one of the following 7 divisions: (1) literature,
philosophy, and history; (2) law and politics; (3) economics and
commerce; (4) fundamental science; (5) engineering; (6) agriculture;
and (7) medicine, dentistry, and pharmacy. Each division would consist of 30 members. Furthermore, the Japan Academy would be set up
within the Science Council, the NRC dissolved, and government
subsidies to the JSPS, a private foundation, would probably cease. It
was hoped that the JSPS could be used as an outside organization of
the Science Council and entrusted with appropriate activities.[79] The
election of members was held in December 1948. The chemist Naoto
Kameyama of Tokyo University and Yoshio Nishina were elected
president and second vice-president, respectively, at the inaugural
meeting, which was convened on January 20, 1949.[80] According to
SCAP opinion, approximately 40 of the 210 elected representatives
were known to be communists or communist sympathizers.[81]

The Science Council of Japan provided a framework for scientist
representation in government policy-making. Its effectiveness was
constrained by the framework negotiated with SCAP and the Japanese
government. Despite the appeals for the "democratization" of Japan,
other concerns also came into play. The Science Council enabled ease
of surveillance and "central control" of scientists, while at the same
time avoided concentrating power in the hands of scientists themselves. As official memoranda show, Kelly and O'Brien made sure that
the "interests of Korea have been considered."[82]

The Scientific and Technical Division of ESS considered the
reorganization of Japan's national scientific organizations as its most

important accomplishment. Unlike progressive Japanese scientists, the Allied representatives viewed science and technology not so much as a means of achieving the democratization of Japan as a means by which to solve industrial problems and achieve economic growth. Related to this were measures aimed at the improvement of quality control, enforcing industrial standards, and improving conditions relating to patents and trademarks. Japan was thus encouraged to carry out research and development in areas that might contribute to a "peaceful and productive nation." Meanwhile, maintaining prohibitions on weapons research, the Scientific and Technical Division carried out a surveillance function by reporting on individual scientists and their work, as well as carrying out inspections of laboratories.[83]

IV   NISHINA'S POSTWAR ROLE

Nishina shared the concern of physicists in ESS that science should assist Japan's postwar reconstruction. After World War II and until the time of his death, Nishina was one of the most active scientists in the reorganization of science in Japan, and, particularly, that of Riken. Former loyalties did not stand in the way. Nishina was responsible for dismissing Nagaoka (his former patron) as chief researcher in early 1947, as the institute was overstaffed and its economic viability was being threatened. Nishina was pragmatic to the end.[84]

Although the war was over, Nishina faced an even greater battle in helping Japan rebuild. He emphasized that the economy should take priority over the pursuit of basic research.[85] In July 1946, as part of the dissolution of the *zaibatsu*, Riken was reorganized. The two physicists Kelly and Nishina went to great lengths to preserve Riken in order to make science work for the nation.[86] Nishina was elected to represent Riken, became its director, and in November 1946, was chosen as its new head. His long years of study overseas had equipped him with adequate English, which helped him deal with the GHQ. As we saw with Rabi, Nishina's international network of scientific contacts would work very much in his favor. He enjoyed a special rapport with the authorities, which placed him in a good position to negotiate the future of science. He believed it was now more than ever necessary to link science with industry for Japan's reconstruction and for Riken's survival. The production of penicillin was one way in which this was done.[87]

By 1948, SCAP was able to distinguish the difference between building A-bombs, and the use of radium and radioisotopes for medical research. Nishina was granted permission to prepare and use stable isotopes for studies to benefit primary industry and medicine. In August 1948, Nishina was made a fellow of the Japan Academy. Around this time, he became resigned to the fact that he would now have to be even more entrepreneurial. In a letter to Bohr that month, he explained:

> The present objective of our Institute is the application of science to peaceful industry in order to promote the rehabilitation of general economy of this country, in which the poverty paralyses its whole machinery. Scientists must take their due share in realizing the economical recovery of Japan in order that she can assume her responsibility in promoting world peace.[88]

Nishina's world had changed and his role and aspirations for science were transformed as well. As he explained to Lawrence,

> In my opinion scientist's way of thinking must prevail throughout the world, if the progress of international politics towards enduring peace is to be achieved.[89]

This echoed the thoughts of physicists abroad. Science and democracy were considered two sides of the same coin. Like the Japanese physicists, even Robert Oppenheimer considered that perhaps

> there are elements in the way of life of the scientist which . . . have hope in them for bringing dignity and courage and serenity to other men. . . . [science enjoys] a total lack of authoritarianism . . . accomplished by one of the most exacting of intellectual disciplines. [The scientist] learns the possibility of error very early. He learns that there are ways to correct his mistakes; he learns the futility of trying to conceal them . . .[90]

Sagane also corresponded with Lawrence:

> I am extremely happy to be able to send my first letter after the war to my beloved teacher Professor Lawrence. As Louis Pasteur states "There is no national boundary in science and scientists have but one country."[91]

The language used in the postwar correspondence from Sagane and Nishina to Lawrence reflect the keywords of the period.

> After all everything which we experience is to build a peace loving, civilized nation so that you may have a better neighbour than ever

before and the ocean between us may be literally "Pacific." In this respect I should much appreciate the future cooperation and assistance from the side of your scientists.[92]

Peace, internationalism, and the authority of the scientific method would be constantly repeated in a variety of contexts over the next two decades. Kelly encouraged interaction between Japanese and American scientists for he felt that the "internationalism" of science would teach the Japanese the benefits of Democracy with a capital "D" (or at least the benefits of a strong U.S.–Japan relationship). As he wrote to Lawrence,

> Your letters [to Sagane and Nishina] have had a very encouraging effect and such actions help us illustrate to the Japanese what we mean by Democracy.[93]

In an election in December 1948, Nishina was elected to the Science Council. In January 1949, he was chosen vice-president of the Council, and in March, he became acting director of the Scientific and Technical Administration Committee (STAC), created to act as a liaison between the Government and the Science Council. In September 1949, he was allowed to attend an international conference, the general meeting of the International Council of Scientific Unions (ICSU) in Copenhagen (September 14–16), as a representative of the Science Council. After leaving Copenhagen, Nishina went on to attend the General Assembly of UNESCO (United Nations Educational, Scientific, and Cultural Organization) in Paris in his capacity as president of the Federation of UNESCO Cooperative Associations in Japan. Nishina was the first postwar Japanese delegate to attend such international meetings. Bohr had instigated the invitation for Nishina to attend the ICSU conference. When the ICSU was unable to cover his expenses, the GHQ bore the cost.[94]

Nishina was able to meet Lawrence and see his laboratory for the first time in early 1950. Upon returning to Japan, he was kept busy with paperwork and meetings.

> I have entirely been taken up with the administration of my Institute and various Government and private committees. I may say I have no free time of my own. Since last June I have been working as Acting Chairman for the Foreign Investment Commission, which deals with the introduction into Japanese industry of foreign technology as investment in kind. This will help Japan build up its industry for the rehabilitation of [the] general economy of the country.[95]

Such commissions were an important part of industrial policy in Japan. In addition, reserves of foreign currency were allocated to Japanese firms to enable them to purchase capital equipment and raw materials, as foreign borrowing was prohibited.[96] Nishina, ever the politician, ended the letter with a postscript:

> The international situation has changed a great deal since I saw you last. We are very glad that a great advance is being made at the Korean war front. We must be very firm against communism.[97]

His disdain for communism was probably prompted by SCAP's "Red Purge" (1949–1950), which cost 20,997 communist sympathizers their jobs in the public and private sectors. In the new Cold War context, they were carefully chosen words.[98]

Nishina, reflecting soberly on the role of scientists in war, wrote around 1948 that

> Scientists are not responsible for starting war, but when war is once started they are called upon to cooperate in its prosecution, and through their cooperation war is made increasingly terrible.
>
> If all scientists in the world unanimously decide not to cooperate in any war, it may go a long way toward preventing war, but such a proposition appears to be well-nigh impossible, as it is supremely difficult in existing circumstances for American scientists to come in contact and act in close concert with Soviet scientists in the matter. It is therefore doubtful whether the well-meant efforts of scientists to prevent war will prove successful. This consideration should not, however, deter them from putting forth earnest endeavors to promote peace.
>
> Scientists in America are collectively carrying on a peace movement. Dr. Einstein is a prominent figure in this movement. Although many American scientists are favorably inclined toward this movement, there are a few scientists who evince little interest in it. Their indifference is due not so much to their positive support of war as to their belief that in the last resort war is unavoidable.[99]

Nishina's pessimistic words reflect the emergence of the Cold War and the inevitability of another global conflict. While the scientists' peace movement brought some hope, Nishina knew that the events were beyond their control. In 1949, he prophetically wrote that atomic energy was not the sole domain of physicists, suggesting that their authority as nuclear experts would not last indefinitely.

> Although it may have its origins in physics, the development of atomic energy has combined physics with chemistry, engineering and other

technologies, and required promotion by the government. In order to achieve a great deal today, one must often rely on the cooperation of many different fields. Up until now, one has been able to work within the one field, but that is now a thing of the past.[100]

In late November 1950, Nishina was taken ill with liver cancer and died on January 10, 1951.[101] Although the Occupation had yet to come to an end, forces had already shaped Japan's postwar agenda. The politician Yasuhiro Nakasone would be a key actor in the years to come.

V   NAKASONE'S POSTWAR ROLE

Yasuhiro Nakasone (b. 1918) rejoined the Home Ministry after the war, and held positions such as section chief of the Kagawa Prefectural Police and the inspector for the Metropolitan Police Department (Keishichō). During the war, the Home Ministry had overseen the repressive activities of the Special Higher Police (Tokkō). Due to such activities, the Home Ministry was dissolved as part of the postwar cleansing process of democratization reforms.[102]

In addition to such changes, SCAP also encouraged the Japanese to form democratic and representative institutions. Leftists established the Japan Socialist Party and Japan Communist Party in November 1945. The conservatives organized themselves as well, with the creation of the Japan Liberal Party and Japan Progressive Party that month. The Japan Cooperation Party was established the following month.[103] It was around this time of political unrest that Jōji Yabe, one of Nakasone's old teachers, encouraged him to return home to Takasaki, a former castle town in Gunma prefecture, where he was born. Yabe feared a communist takeover of the town by left wing workers, especially those belonging to the national railway. In order to "save" his town, Nakasone returned in late 1946 to stand for the 1947 Diet election as a Minshutō (Democratic Party) candidate.[104] The Democratic Party (formally established in March 1947) had evolved from the Japan Progressive Party.[105]

As part of his election strategy, Nakasone formed the "Seiun Juku" (literally translated as "Blue Cloud School" or "Lofty Aspirations School") in Takasaki in 1947. The aim was to bring together a group of young men, with a similar (anticommunist) political ideology, who would support him. This school resolved that

> We shall build a national community and speedily restore our dignity as an independent nation. We shall construct an Asian-style democracy.

Aspiring to the civilization of a new age of creativity, we shall reform the institutions of the world and of our homeland.[106]

Nakasone's words encapsulated his approach to the reconstruction of Japan. Japan would be reconstructed with the help of science and technology and would become a more creative nation, under a democracy that would be different from that of the West, and unlike the democracy as espoused by left wing Japanese physicists such as Shōichi Sakata and Mituo Taketani. Before and during the war, there had been moves to integrate the government, the military, and industrial production. Nakasone was a conservative nationalist who felt that the people should again be mobilized for Japan's reconstruction.

Nakasone's family owned a lumbering business and were well known in the district around Takasaki. This helped in his quest to transform himself from a bureaucrat to a politician. Equipped with a charcoal-burning truck, probably courtesy of his family, and a white bicycle, Nakasone personally did the rounds of the electorate, canvassing voters. This was unusual at the time, such tasks normally being relegated to campaign supporters. Nakasone's white bicycle was likened to a white horse from the Emperor's stable,[107] both giving the rider delusions of grandeur. Apparently, "Nakasone was one of the first politicians to build a political organization based directly on the people."[108] This skill of being able to mobilize talented people would be put to great use throughout his career. All this combined to give Nakasone his first victory in the elections at the young age of 28.

Elections under the new constitution gave the Japanese a taste of American-style democracy. American planning for the occupation of Japan, even before the end of World War II, generally reflected the social democratic philosophies espoused by Roosevelt's "New Deal." Japan would be reformed and demilitarized with an agenda that included de-concentration of economic power, land reforms, and the breaking-up of the military–industrial complex. It is generally accepted, however, that by late 1947, by which time Nakasone had been elected to the Diet, American leaders increasingly saw Japan, rather than the Chinese Nationalist Government, as the means by which to contain the spread of communism in Asia.[109] This tended to concur with Nakasone's anticommunist and militaristic vision for Japan, but many Japanese and "New Dealers" in SCAP felt otherwise.

It can be pointed out, however, that former president Herbert Hoover had urged Truman, who had become president on April 12, 1945, to consider both Japan and Germany as potential allies against the Soviet Union from as early as that very year. Japan should be dealt with kindly and allowed to retain its colonies in Korea and Taiwan. Hoover favored the reconstruction of Japan as a bulwark against communism from the very beginning of the Occupation.[110] Maj. Gen. Charles A. Willoughby, who headed the Intelligence Section, G-2, of SCAP, shared Hoover's view that the fight against communism should take priority over the democratization of Japan. As soon as three days after the Occupation had begun, Willoughby had begun recruiting high-ranking Japanese officers for his organization. One group of Japanese officers was involved in activities against the Japan Communist Party. Willoughby was also involved in arranging immunity for Japanese scientists involved in chemical and biological warfare research.[111] It appears that both Prime Minister Shigeru Yoshida and Gen. Douglas MacArthur were unaware of many of these activities.

Nakasone was first elected to the House of Representatives in April 1947, the first such election under the new "MacArthur Constitution" or "Peace Constitution." The Japan Socialist Party took the reins of a coalition government for the only time in Japan since the end of the war. The Tetsu-Katayama-led cabinet collapsed in February 1948 when it failed to have its proposed budget approved. Another coalition government led by Suehiro Nishio took office, only to be ousted in October 1948 because of a corruption scandal involving the Showa Denkō company.[112]

Despite liberal reforms and a socialist-led government, it is generally agreed that there was a substantial change in Occupation policy during the period 1947–1948. The push for democratization and demilitarization gave way to a desire to establish Japan as a solid ally of the United States. This change in direction has sometimes been called the "reverse course." The Chinese communist revolution in 1949 and the Korean War in 1950 further reinforced the U.S. commitment to achieving social and political "stability" in Japan.[113]

During Nakasone's first two decades in parliament, he was known as an aggressive and ambitious politician who was straightforward in his actions.[114] In January 1949, he was elected for the second time to the House of Representatives, along with the third Shigeru Yoshida government. He became allied with the conservative side opposing Yoshida's administration. Unfortunately for Nakasone, Yoshida was prime minister for much of the period between 1946 and 1954. Nakasone attacked

Yoshida's emphasis on cooperation with the United States as the humiliating diplomacy of a lowly dependent, the young Turks [of whom he was one] advocated an "independent constitution" to replace the document drawn up during the occupation. They also favored the remilitarization of Japan and censured the complacency and corruption of Yoshida's autocratic rule. Senior statesman Matsumura Kenzo compared the Nakasone of this period to a "young warrior riding high in the saddle."[115]

During the Occupation period, Nakasone is said to have attended the Diet parliament wearing mourning attire complete with black tie, as an indication of his feelings of sadness and indignation at the policies and reforms that were implemented.[116]

In March 1950, Nakasone, the "young warrior," became a member of the preparatory committee for the formation of a National Democratic Party.[117] Around this time, the 32-year-old Nakasone was responsible for a 7000-word petition to Gen. Douglas MacArthur, which requested a revision of the constitution and asked for the formation of an independent defense force for Japan. Although MacArthur is said to have paid the petition little heed, Nakasone persevered with his hopes for a rearmed Japan and went as far as writing a song entitled "The Revision of the Constitution."[118] Neither the foreigner-imposed constitution nor the changes in Japanese values, which resulted from the largely American-led Occupation, was to his liking. In June 1950, Nakasone went to Europe to attend the world conference of an association aimed at the restoration of moral values in society. He took this as an opportunity to visit the United States and various countries in Europe as well, for such a chance to travel overseas was still rare during the Occupation.[119]

Japan's rearmament can be considered as being triggered by Truman's decision to intervene in the Korean civil war. In April 1950, John Foster Dulles was appointed top adviser to the secretary of state Dean Acheson, for Asian matters, and especially the peace treaty with Japan.[120] Dulles arrived in Tokyo in June 1950, urging Prime Minister Yoshida to greatly expand Japan's armed forces, in addition to allowing U.S. bases to remain in Japan.[121] The Yoshida government was ordered by MacArthur in July 1950 to establish a Police Reserve Force consisting of 75,000 men. The Maritime Safety Force would also be increased by 8,500 men.[122] In August 1950, the Police Reserve Force was formed and this later became the Peace Preservation Agency on August 1, 1952. That very same day, the Peace Preservation Agency Technical Research Institute was established and with it the recommencement of what some commentators view as military research. That very summer of 1952, the Federation of Economic Organizations (known as "Keidanren" in Japanese) established the Defense Production Committee. The Arms Production Cooperative Association was started and this later became the Japan Arms Industries Association (Nihon Heiki Kōgyōkai). Its aim was to conduct studies and research into the technical side of weapons.

Dulles returned for a further round of talks in January and February 1951. At this time, Nakasone requested a meeting with Dulles in order to ensure that Japan would not, after the signing of the Peace Treaty, be constrained with respect to civil aviation or research into atomic energy, crucial areas, which had been under prohibitions during the Occupation. Without development of these areas, Nakasone feared that Japan would be left to stagnate as an agrarian nation.[123] In September 1951, the Peace Treaty was signed by the United States and Japan, bringing the Occupation and the Pacific War formally to an end in April 1952.[124]

Nakasone was enthusiastic about helping shape party politics at this formative time. He became a member of the preparatory committee for the Kaishintō ("Progressive Party" or "Reform Party"), which included a number of politicians who had been purged at the end of World War II, as well as liberal-minded conservatives. (This party would later join the Japan Democratic Party formed by Hatoyama in 1954.[125]) In October 1952, Nakasone was reelected the third time.

CONCLUSION

In this chapter, we have seen that despite the desire of many after the war to rationalize and "modernize" the organization of science and technology in Japan, longstanding differences in the interests and ideologies of scientists, industry, and government worked against this. The emergence of American-imposed democracy involved many changes, which were more cosmetic than real. Democracy could only be exercised within certain limits and could be altered in response to what might be perceived as a threat to American interests.

The immediate ramifications of the democratic Allied Occupation were to prohibit rather than encourage research and the surveillance of laboratories in a spirit quite alien to the "freedom of science" supposedly cherished by the West. Many scientists had hopes that the Science Council of Japan would become the central organization for science and technology in Japan. Had Nishina survived, this may have come into fruition, but Nakasone's political activities and lobbying ensured that it would be more closely tied to the needs of the state. The next chapter describes how two physicists, Sakata and Taketani, found the promise of democracy a tantalizing opportunity to empower scientific workers to influence their research environment. Like Nishina, they would grapple with the realities imposed by Cold War tensions, which Nakasone's career so clearly reflected.

CHAPTER 4

PHYSICISTS ON THE
LEFT: SAKATA AND TAKETANI

As in wartime Japan, postwar scientists were asked to subordinate their individual freedom to conduct research for the interests of the much larger group—the nation. This was a source of continuing conflict. Despite Nishina's call for science to first help revitalize the Japanese economy, Shōichi Sakata and Mituo Taketani responded by arguing for research autonomy and a commitment to basic research. This chapter focuses on the role of the two physicists, Sakata and Taketani, in the postwar democratization of science and technology. Despite great differences in socioeconomic background (described in the first two sections), both shared a common interest in Marxism, reflecting the times in which they lived. Both were particularly outspoken regarding the need for science and democracy in "modernizing" Japan. They held the view, as many scientists after the war, that science and democracy were matching wheels for social progress.[1] The third and largest section of this chapter describes how Sakata and Taketani attempted to prove the veracity of this conviction.

Taketani would write about this whereas Sakata hoped to actually transform society by democratizing the research system and influencing policy. In their different ways, both physicists promoted scientism, over-valuing science as a potential savior. Their enthusiasm can be linked to the pervasiveness of Marxism in the intellectual world, and also to the technological fatalism and feelings of scientific necessity, which gripped the Japanese at the end of the war. To Marxists, reconstruction was equated with revolution. The social system had to be changed so as to allow science to be used effectively as a major force for social change. But the Supreme Commander for the Allied Powers (SCAP), the Japanese government, and industry thought otherwise.

The golden age of the Marxist intellectual, the setting for this chapter, was a response to the poor conditions that Japan was enduring for the first 15 years after its defeat. After this time, Japan became more accepting of, or resigned to, its relationship with the U.S. economic prosperity, and greater contact with the outside world changed the Japanese view of the future to a more optimistic one, not so much in need of a revolution, and one less reliant on physicists.

I SHŌICHI SAKATA

It was a bright, clear spring day on May 11, 1950. The cherry blossoms had already come and gone in Ueno Park, and the distinguished Tokyo University professor Samurō Kakiuchi and his wife Tazuko (daughter of a famous anthropologist) slowly left the train station and made their way to the Japan Academy. With them was their daughter, Nobuko Sakata, who held the hand of her grandmother, Kimiko Koganei, sister of well-known author Ōgai Mori. Upon arriving at the Academy, they saw the impressive car of Prime Minister Shigeru Yoshida parked outside the building.

At the entrance, a woman dutifully attached name badges to their chests and the small group then proceeded to the first floor waiting room where the 39-year-old physicist, Professor Shōichi Sakata, waited with his parents, former governor Mikita Sakata and his wife Tatsue. Professor Sakata stood out from the other award recipients and their guests for he wore a loud yellow chrysanthemum with a bright red ribbon on his lapel. After a speech from the president of the Academy, Sakata received the Imperial Award in a blaze of flashlights and to the applause of the assembled audience. What brought all these well-connected people together and why would the unassuming Professor Sakata, a short and stocky man with a high voice, stay in the spotlight for the next 20 years until his death?[2]

Like Yoshio Nishina, Shōichi Sakata was born into a family with a tradition of public service. Born on January 18, 1911, the eldest son of Mikita and Tatsue Sakata, his first home was in the rarefied space of the prime minister's residential compound in Tokyo. Mikita was the adopted son of Masaharu Sakata, who fought for the Chōshū clan during the Meiji Restoration and who was elected to the House of Representatives under the new government. He had graduated from the prestigious Faculty of Law at Tokyo Imperial University, and was secretary to Prime Minister Tarō Katsura. In March 1912, he took up a new position in Fukuoka prefecture, but returned to Tokyo as secretary to the minister for commerce and agriculture (1914) and secretary to the minister for home affairs (1915). He later went on to become governor of Ehime (1916) and Kagawa (1917) prefectures, mayor of Takamatsu city in 1919, and retired from politics in 1920.[3] Shōichi, too, was academically gifted, and school provided him with opportunities to steep himself in physics. In 1924, he was fortunate to be taught by the physicist Bunsaku Arakatsu while at Kōnan Middle School in Hyōgo prefecture. He then proceeded to study at Kōnan High School (1926–1929) during the early years of the Shōwa period, a time when the theory of relativity and quantum mechanics were generating a great deal of interest in physics in Japan. His reading matter at the time included books by Jun Ishiwara, the physicist–historian Ayao Kuwaki, and the philosopher Hajime Tanabe. In 1926, while only a high-school student, he had the formative experience of attending a lecture by Ishiwara. And it was at Kōnan High that Sakata became close friends with Tadashi Katō, who, in 1929, translated (with Eijirō Kako) Engels' *Dialectics of Nature*. Katō would later came in contact with a progressive group of intellectuals, based at Kyoto University, who produced a journal called *Sekai bunka* (*World Culture*). Mituo Taketani was a member of this group.[4] It was, thus, through a combination of a privileged background, educational opportunities, and fortuitous friendships that Sakata's identity would be shaped.

Yet, despite his elite background, Sakata's fellow students and teachers could tell from his handwriting that here was someone who was antagonistic to the traditions of Japanese calligraphy, if not its culture and history. Sakata's brushstrokes constituted an array of awkward shapes, which were easy to read despite their ugliness. Unlike the elegant flowing lines of the handwriting of Yukawa or Tomonaga, Sakata's writing betrayed his desire for modernity as opposed to tradition.

After graduating from Kōnan, Sakata became an auditor student in the Department of Physics at Tokyo Imperial University. That year, he visited Yoshio Nishina, who had returned from Copenhagen the year before. Nishina was a distant relative of Sakata (his great-aunt's younger brother-in-law). In April 1930, Sakata entered the Department of Physics at Kyoto Imperial University, where he met Nishina again in 1931 during a special lecture. In 1932, Sakata chose nuclear theory as the subject of his graduation thesis, and in March 1933, graduated from Kyoto Imperial University, some four years after the physicists Yukawa and Tomonaga.[5] Sakata found his way to the physicists' mecca—Nishina's laboratory at Riken—where he was admitted, first as a research student, and then as a research associate for the period April 1933–March 1934.

He worked with Tomonaga and became acquainted with Mituo Taketani, who would be a lifelong friend. They were compatible in many ways. Sakata found that Taketani and he shared similar interests—both had studied at Kyoto University, wore glasses, and were even of similar physical build.

After Riken, Sakata moved to Osaka Imperial University as an associate, where he worked with Yukawa on topics such as the meson theory. In 1938, Sakata was promoted to lecturer, and also obtained a part-time Riken consultancy in 1939. With Yukawa's move to Kyoto Imperial University, Sakata took up a lectureship at Kyoto, in addition to his lecturing responsibilities at Osaka Imperial University.[6]

On December 6, 1939, Sakata married Nobuko Kakiuchi, second daughter of Samurō Kakiuchi, a Tokyo University professor and founder of Japan's *Journal of Biochemistry*.[7] Her mother, Tazuko, was the elder daughter of the famous anthropologist and professor of anatomy, Yoshikiyo Koganei. Koganei,[8] in turn, was of samurai background and married to Kimiko, the younger sister of the Army medical affairs bureau chief and author, Rintarō Mori (also known as Ōgai Mori).[9] Such family connections may not have directly assisted Sakata's career, but they certainly reinforced his already impressive lineage. Nobuko created a home environment, which provided important support for his more public activities and academic interests.[10]

In February 1940, Sakata became a member of the Physics Research Committee of the National Research Council (NRC), thus, getting his first taste at policy-making. In May 1941, Sakata received his Doctor of Science and in April 1942, Yasutaka Tanikawa and he presented the two-meson theory at Riken. Later that year, Sakata moved to Nagoya to become professor at the new imperial university there. In January 1943, perhaps as part of the mobilization of science, Sakata was made a part-time member of the Institute for Chemical Research at Kyoto Imperial University. In March 1945, with the increased bombing of Japanese cities, Sakata's wife and children took refuge in the Kakiuchi family holiday home at Lake Yamanaka near Mt. Fuji. In April, the Nagoya theoretical physics group moved to Fushimi in Nagano prefecture, using the premises of the Fushimi People's School (Fushimi Kokumin Gakkō). Taketani was invited by Sakata to Fushimi to discuss physics.[11]

During his spare time at Fushimi, Sakata read J.D. Bernal's *The Social Function of Science* (1939). He would eventually co-translate the book into Japanese in 1951, along with Yoshirō Hoshino and Makoto Tatsuoka.[12] The book was a revelation to Sakata, and seemed to be a natural extension of Marxist thought, which he had been introduced to while at high school. Bernal likened science to communism:

> In its endeavour, science is communism. In science men have learned consciously to subordinate themselves to a common purpose without losing the individuality of their achievements. . . . In science men collaborate not because they are forced to by superior authority or because they blindly follow some chosen leader, but because they realize that only in this willing collaboration can each man find his goal. Not orders, but advice, determines action.[13]

Such words served to reinforce the authority of scientists and cast doubts on whether science could prosper under capitalism. Sakata was particularly impressed by Bernal's concept of "laboratory democracy," which combined "personal direction with democratic control."[14] This seemed to be a way of overcoming the "feudalistic" domination of researchers by professors in Japan. Sakata strived to ensure that those same freedoms would be enjoyed by all scientific workers. Sakata's "democracy" embraced not only elite physicists but included representation and participation by a much wider group of people. Bernal helped set an agenda for Sakata, which would last the next 25 years— a "science policy" agenda that emphasized basic over applied research, and the need for regulatory policy, which policed the potentially harmful effects of science.

II MITUO TAKETANI

It was not a happy new year for the physicist Mituo Taketani in 1939. Detained in a Kyoto police station for interrogation, along with a cell full of gamblers, pickpockets, thieves, and robbers, there was little room to sleep and little joy in what the coming year might bring. What brought Taketani to such unhappy surroundings?[15]

Unlike Sakata and Nishina, Taketani came from a more humble background. He was born in 1911 in the city of Ōmuta in Fukuoka prefecture, the third son of a technical expert at the local Mitsui-owned Miike coal mine. When Taketani was six, the family moved to Osaka. Taketani's father found employment as an engineer at a factory owned by an uncle, which was located close to Osaka. The end of World War I saw a decline in the fortunes of the factory and Taketani's father was forced to find a new job at a coalmine in the northern part of Taiwan. After completing the first year of primary school in Japan, Taketani went to Taiwan and continued his education at a Japanese school there. Taketani's father subscribed to popular science magazines such as *Kagaku chishiki* (*Scientific Knowledge*) and *Kagaku gahō* (*Science Illustrated*). These seem to have helped trigger an interest in science in the young Taketani. His experiences in Taiwan also provided him with firsthand knowledge of the injustices of Japanese colonialism.[16]

Taketani gained admission to the Department of Geophysics in the Faculty of Science at Kyoto University. In his second year of university, he switched to a physics major. That same year, he also began attending lectures in the Faculty of Arts, where he had the opportunity to hear the famous philosopher Hajime Tanabe speak. Amongst Taketani's friends at Kyoto were Mitsuo Hara in the Department of Chemistry and Hakuo Mita in the Department of Mathematics. Both Hara and Mita were influenced by the activities of the Tokyo-based Society for the Study of Materialism (Yuibutsuron Kenkyūkai) and the natural dialectics of Engels. Jun Tosaka was a central member of the Society.[17]

In 1933, around the time of Taketani's third year at Kyoto University, the Takikawa Incident (sometimes described as the "Takigawa Incident") occurred wherein the minister for home affairs demanded the dismissal of law professor Yukitoki Takikawa because of his alleged communist sympathies. Many academics and students protested against this government interference in the universities.[18] Despite concern showed by students in the Faculty of Science, action amongst academic staff tended to be confined to the Faculty of Law.[19]

Taketani graduated in 1934, his final year project being a broad thesis entitled "How should we study nuclear physics?" This was revised and later published in the journal *Sekai bunka* (April 1936). The message was that we should pursue nuclear physics according to a three-stage theory of cognition. Around this time, Taketani obtained a small income by teaching physics part-time at a middle school in Kyoto, while also occupying an unpaid position as an assistant in the Yukawa laboratory at Osaka University, where he joined Yukawa, Sakata, and later Minoru Kobayashi working on the meson theory. In late 1934, shortly after graduating, Taketani was invited to become involved in an intellectual group, which would publish a revamped version of *Bi hihyō* (*Art Criticism*) entitled *Sekai bunka*. The aim of the magazine was to protect antifascist culture. The first issue of *Sekai bunka* came out in February 1935. Original papers by various people, especially specialists in philosophy, aesthetics, literature, and the arts, were included, along with reports of developments in the antifascist movement abroad. Taketani wrote under the pen name of Kazuo Tani. The print run was a substantial one of about 1000 copies.[20]

In the autumn of 1936, however, half the core members were rounded up by the Special Higher Police[21] and interrogated. Taketani busied himself in helping to develop Yukawa's meson theory in the period 1937–1938. In January 1938, Taketani decided that it was safer to leave Kyoto to stay with a friend in Kobe. In June, a second round of interrogations occurred. On September 13, Taketani was taken away by the police on the same day that a communist group was interrogated. He was taken to Kyoto and detained, to await investigation. He was made to sign a statement that said *Sekai bunka* supported the Japanese Communist Party and was kept in detention till April 1939, when he was released with Yukawa acting as guarantor.[22]

After his release, Taketani worked in the Osaka University laboratory of Seishi Kikuchi, as an unpaid associate. His research included theoretical work as well as cyclotron experiments. The income came from the money Kikuchi received from Riken and from teaching at a school attached to the Fuji Nagata Shipyard. In the summer of 1940, Taketani was hospitalized for one month as he was suffering from pleurisy. Physicist colleagues were concerned about Taketani's plight and invited him to come and help Sin-itirō Tomonaga in Nishina's laboratory from around April 1941. His status was, again, that of an unpaid research associate. It was arranged, however, that Taketani would receive a scholarship from the publisher Shigeo Iwanami from the beginning of that year. This was augmented by another scholarship from the Hattori Hōkōkai.[23]

The Nishina lab by this time included a cosmic-ray experiment group, a cyclotron group, a theory group, and a biology group. The cosmic-ray group included Yatarō Sekido and Masa Takeuchi among others. Tomonaga was one of the leading members of the theory group and Ryōkichi Sagane was the core member of the group working on the cyclotron. Fission studies were also carried out at the lab by physicists such as Kenjirō Kimura. The Army-funded atomic bomb project began in 1943 and Taketani was mobilized for the project, along with Masa Takeuchi, Kunihiko Kigoshi, and the cyclotron group. The aim was to separate uranium 235 by gaseous thermal diffusion method. Nishina considered this the most practical method of isotope separation, perhaps because thermal diffusion had been used by Eiichi Takeda at Kikuchi's laboratory at Osaka University.[24] Taketani assisted with theory and Hidehiko Tamaki was the overall coordinator. Taketani married a Russian woman medical student at the beginning of 1944. His work and new marriage were interrupted in May 1944 when the Special Higher Police detained him for a second time. His marriage probably made him even more suspect of subversive activities than before.[25]

Taketani made it known to the police that he was working on an important Army project. Visitors such as Satoshi Watanabe from the Nishina lab also impressed this on the police. During his detention, Taketani worked on theory relating to the thermal diffusion process and also calculated the potential damage from an atomic bomb. A bad bout of asthma enabled him to return home on September 10, 1944, with the proviso that he regularly report to the police station.[26]

At the end of the war, Taketani's mother and eldest brother returned to Japan from Taiwan. His father had died there around the time of Taketani's first period of detention. Taketani found some employment as a teacher at the Tōkai Science College, which was attached to Tōkai University. The College had formerly been called the Denpa Kagaku Senmon Gakkō (Radio Science College) with Nishina as the principal. During the war, Taketani had taught there on and off in between detentions. The end of the war saw Taketani become one of the four most well-known physicists, along with Yukawa, Tomonaga, and Sakata. Unlike the latter three, Taketani was respected less for his achievements in physics than for his role as a social spokesperson, promoting the activities of the Elementary Particle Theory Group, and commenting on issues relating to the social responsibility of scientists.[27]

III Postwar Democratization of Science and Technology

Immediately after the end of the war, Japanese from both ends of the political spectrum called for the mobilization of Japanese science and technology. The debate tended to focus on how science would be used and for what purposes. Left wing activists emphasized the need for a democratic administrative structure and argued for the improved welfare of the people. Socialism was portrayed as a type of scientific remedy to solve Japan's problems. This attracted scientist–intellectuals such as Sakata and Taketani who saw it as a means of arguing for the primacy of science, the social responsibility of science, and to protect science from any abuse by the state or private sector. Sakata felt, like Bernal, that society would need to change for science to prosper. In the first couple of years of the Occupation, a socialist revolution seemed a possibility.

On January 12, 1946, around 200 scholars from the social sciences and natural sciences gathered in the auditorium of the headquarters of the Japanese Red Cross in Tokyo, to attend the inaugural meeting of the Democratic Scientists' Association (abbreviated in Japanese as "Minka"). Minka has parallels with the Association of Scientific Workers in England and the Federation of Atomic Scientists in the United States[28] It aimed to carry on from where the prewar Proletariat Science Movement had left off by (1) building a "democratic" science; (2) resisting feudalism and militarism in science; (3) obtaining freedom in scientific research and its publication; (4) fighting the antidemocratic education system and policies; (5) democratization of scientific facilities and organization; (6) having science and technology contribute to the enrichment of the lives of Japanese people; (7) the training and encouragement of up-and-coming scientists to help build a democratic Japan; (8) encouragement of cooperation and dialogue between natural scientists, social scientists, and technologists; (9) support for the social role of scientists and technologists; (10) national and international liaison between democratic science and technology groups; and (11) facilitation of the absorption of scientific findings of democratic countries.[29] The agenda was an attempt by Japanese scholars to forge a new identity for themselves by severing any ties to their feudalistic past and reshaping existing institutions.

The membership of Minka included members of the former Group for the Study of Materialism (Yuibutsuron Kenkyūkai) (which had promoted a proletarian science), materialists, Marxists, democratically minded scholars, and students. It was hoped that a network of branches would be built throughout Japan. At the height of its popularity in late 1949, it had 110 branch offices, a specialist membership of 2000, and a student and nonspecialist membership of 11,000. This was a large-scale movement of a type never before seen in Japan. Its president was the historian of science Kinnosuke Ogura, and the numerous permanent secretaries included Mituo Taketani and Shōichi Sakata.[30] At its inception, the membership register of Minka also included the names of Sin-itirō Tomonaga and Hideki Yukawa. Yukawa's membership was marked as being under negotiation.[31] Although none of these physicists are known to have joined the Communist Party, the Occupation authorities considered Minka a front for communists.[32]

An indication of the size and influence of Minka is the list of periodicals it produced: *Shakai kagaku* (*Social Science*), *Shizen kagaku* (*Natural Science*), *Riron* (*Theory*), *Rekishi hyōron* (*Historical Critiques*), *Geijutsu kenkyū* (*Studies in the Fine Arts*), *Seibutsu kagaku* (*Biological Science*), *Ningen kaihō* (*Liberation of Mankind*), *Kagaku nenkan* (*Science Yearbook*), *Nōgyo nenpō* (*Agricultural Annual Report*), and *Nōgyō riron* (*Agricultural Theory*). The magazines *Kagakusha* (*Scientists*) and *Gakujutsu tsūshin* (*Science News*) appeared three and two times a month, respectively.[33]

Minka's publications were part of the "literary renaissance" soon after Japan's surrender, in which a veritable explosion of "sōgō" (general) magazines advocated progressive ideas, backed by Marxist intellectuals reveling in their newfound freedom. The SCAP Civil Censorship Detachment (CCD) limited the freedom of writers, but it can be argued that prewar censorship was much more restrictive.[34] There are estimates that between 1946 and 1949, 110 serious literary magazines were being published. Despite the challenge of finding enough to eat from day to day, the hunger for reading and intellectual stimulation was voracious.[35]

Notable magazines were *Kaizō* (*Reconstruction*), *Sekai hyōron* (*World Review*), *Shinsei* (*New Life*), and *Shisō no kagaku* (*The Science of Ideas*). Taketani's writings through such magazines had a major impact on the Japanese intellectual world. They were an important way of engaging with the public and nurturing an image of science that suggested that science was helping to define a new Japanese political order. Marxism was seen as a science. There was, as Bruno Latour puts it, "a translation of political terms into scientific terms and vice versa."[36] An essay on Taketani's Marxist-inspired theory of technology[37] appeared in the February 1946 edition of *Shinsei*. Taketani considered technology as the application of objective laws in productive, human practice. He distinguished between technology and skill, the former is objective and enriched by the bank of knowledge whereas the latter is subjective and acquired through experience. Both feed into each other in a dialectical process. New technology may require new skills. When skills are transformed into technology, production capacity and quality often improve.

In the first issue of *Shisō no kagaku* (May 1946) Taketani wrote a critique of contemporary philosophy. An article entitled "Modern Physics and Theories of Cognition" appeared in the July 1946 edition of the magazine of the natural sciences section of Minka. "Galileo's Study of Dynamics" was published in *Kagaku* (*Science*) (April 1946) and "The Atomic Age" was included in *Nihon hyōron* (*Japan Review*) around 1947. Other writings soon appeared in *Shisō* (*Ideas*), *Shakai taimusu* (*Social Times*), *Kyōiku shinbun* (*Education News*), *Tosho* (*Books*), and *Fujin kōron* (*Women's View*). After the Tōkai Science College ceased operating, Taketani was able to make a living by such writing.[38] His extensive writings ensured that his knowledge of physics be seen as relevant to public affairs.

Taketani was one of the founding members of the intellectual group centered around the magazine *Shisō no kagaku*. Others included Shunsuke Tsurumi, Kazuko Tsurumi, and the historian Masao Maruyama. Taketani also participated in some activities of the Japanese Communist Party. He contributed to the establishment of a science and technology section, which he and others wrote about in *Zen-ei* (*Vanguard*) (nos 10–11, November 1946).[39] Concerned perhaps by communist activity, the Economic and Scientific Section (ESS) of SCAP requested a full translation of the article entitled "Shortcomings of Science and Technology in Japan and the Duty of the Communists," which they eventually received in May 1949. The article asserted that:

> Japan will never be able to rebuild herself basically unless she extermi-nates her feudalistic administrative structure and establishes a popular republic. The role which must be assumed by science and technology in the political and economic reconstruction is extremely significant. The overcoming of the current destitution is directly connected with the reconstruction of a new Japan. Systematic mobilization of science and technology is necessary in order to realize the above to any extent. Without the maximum application and development of science and technology, it will be impossible to realize a democratic revolution and the subsequent establishment of a socialistic society, not to mention the fulfillment of urgent reconstruction plans (increased coal output, allevi-ation of housing and food problems and national land-planning).[40]

A less "feudal" and a more scientifically rigorous science and technol-ogy was deemed to be necessary. A communist checklist of the weak points of Japanese science and technology listed the following: the "colonial" nature of technology; the impracticality of science; the uneven development of science and technology; the unscientific nature of technology; the adverse effect of science and technology on the people; lack of scientific methodology; poor agricultural tech-niques; and the feudalistic and bureaucratic nature of universities and public experimental and research institutes.[41] In Sakata's opinion, the Japanese lack of creativity in basic science was not a result of national character but more of environment. It was time to work for an envi-ronment that was congenial to research of a fundamental nature.[42] Thus physicists' own analyses of their history and the failings of Japanese science informed the debate on the role of science in Japan's reconstruction.

A strong concern of Taketani and the communists was the neglect of the safety and welfare of workers. It was felt that industrialists had been and were more concerned with maintaining a supply of low-wage labor and using imported technology. As a result, "research facilities have only a secondary and ornamental significance. Therefore, research funds and budgets are extremely meager and the living standards of the researchers are very low."[43]

Taketani made his thoughts known on the failings of Japan by direct representations to the GHQ, especially through the CIS (Civil Intelligence Section). He felt a strong need to call for the immediate release of political prisoners, the end of feudalism through land reforms, and the scrapping of the Ministry of Education (which had been active in wartime indoctrination). In terms of reconstructing Japan, he wrote in his "An Analysis of Japanese Technology and the Rebuilding of Industry," *Gijutsu* (*Technology*) (March 1946) that there should be a proper place for unions in the process. Increasing the lot of the workers should be part and parcel of improvements to technology.[44] In the same issue, he likened Japan's plight to that of the USSR:

> The present wretched state was once experienced by Russia after World War I, and was surmounted by the leadership of laborers. Though our situations are very different from that of the Russian Revolution, we can learn many items on the Russian crisis. The role of technology in the construction of USSR we can know from some books published. In USSR cooperation of technologists to the socialistic construction, though it was very difficult, was carried on with success, while the technologist could find their true position, and make their effort in the national welfare, proving their importance in the settlement of the difficulties in national economy.
>
> Contrarily the situation is different in the present Japan. The answer of some influential technologist to my ask [*sic*] in the necessity of research could be summarized as follows: science and technology relate only to the war and not to the national welfare.[45]

Taketani's theorizing about technology also called for solutions to problems in social structure, workplace safety, and prevention of accidents. He felt that the Japanese should work to change their perception of technology. Technology tended to be associated with the war effort, the only industries considered "peaceful" and untainted were textiles,[46] even though they were used in military uniforms and bomb balloons during the war.

These and other writings helped Taketani fashion himself as a left wing intellectual; his social power derived not from mere intellect and connections, but relied substantially on media representations.

Ironically, Taketani's claim to fame, his postwar social dominance as a scientist–activist, in no small part owed much to his wartime experience of oppression under the Special Higher Police. It showed that he was the "real thing."

On April 10, 1946, the first general election since the end of the war was held. The perceived importance of science and technology was such that each political party proposed policies for science and technology, which were, in turn, commented upon and criticized by scientists.[47] On April 30, 1946, around 100 physicists gathered at the University of Tokyo to celebrate the inception of the Physical Society of Japan.[48] Besides several reports on the "democratization" of science and technology, Sakata spoke on democratization in the Department of Physics at Nagoya University. He had been inspired by Bernal who argued that

Science can never be administered as part of a civil service, but recent developments both here and abroad, particularly in the U.S.S.R., point to the possibility of combining freedom and efficiency in scientific organization.[49]

He suggested that the organization of research could be tackled by examining how laboratories should be run and how they could be coordinated. But he warned that

no scheme of extensive organization of science . . . will be of the slightest value if the individual laboratories are run in any way inimical to the fullest and freest development of each individual scientist's work.[50]

The moves to establish a Science Council sought to improve liaison amongst laboratories and different fields of scholarship. The question of whether control of future joint-use research institutes should be exercised through a board of scientists, or through the government, would be a major point of debate over the next two decades. In the organization of an actual laboratory, Sakata felt that two basic principles were necessary: (1) freedom for each individual researcher; and (2) effective cooperation in research. For Sakata, democracy also meant autonomy, the freedom to engage in research of one's choosing without fear of censorship. Sakata devised a Laboratory Council for liaison and deliberations by laboratory groups contained within a department, at which all researchers had a voice in policy. This was intended to overcome the previous dictatorial authority of the professor. Research themes, rather than the actual laboratories, would become the basic unit within a department. The Department Council would be the highest deliberative organ within the department, and would look into matters of research and education through the Research and Education Committees, where student representatives as well as staff could participate. On January 24, 1946, the Nagoya University elementary particle theory research group held the first meeting of the Laboratory Council according to these ideas.[51] These principles provided the basis of the Constitution of the Nagoya University Department of Physics, which was put into effect from June 13, 1946.[52]

Sakata's innovations in department organization influenced universities throughout Japan, but met with strong resistance from professors who saw in his plan a threat to their traditional authority. Sakata and Bernal felt that

> Science . . . is itself a democracy: one always open to conviction but not accepting any dictum until it has been convinced. In so far as science infuses government, it enhances all the democratic elements in it.[53]

Sakata tried to extend laboratory democracy beyond the walls of his university department. In an article in the *Tokyo Imperial University Newspaper* (no. 1004, Wednesday, November 13, 1946), Sakata called for the abolition of sectionalism in institutions and democratization of the research system:

> Researchers are completely absorbed in an Ivory Tower secluded from the community and each laboratory is in the state of isolation without mutual help. . . . The everlasting development of science prospers only in the nation who is desirous for the emancipation of her people. In this sense a scientist cannot be allowed to take a neutral attitude in political problem [*sic*]. . . . It is not possible to separate the democratization of national structure and the democratization of research organs. . . . "Laboratory Democracy" is an only effective means for the prevention of vicious growth of feudalism, . . . It must be extended not only to all the scientific organizations but also to the development of a whole community, if it were to be brought into real value. And only then it will become the basis of the unlimited development of science.[54]

Despite the promise of democratization during the Allied Occupation, the early years saw a high level of surveillance of all types of communication in the hope that "There shall be no destructive criticism of the Allied Forces of Occupation and nothing which might invite mistrust or resentment of those troops." Furthermore,

"Nothing shall be printed which might, directly or by inference, disturb the public tranquility."[55]

During the Occupation, the CIS's CCD reviewed material for publication, deleting or making additions to material. Such censorship activities assisted SCAP in its intelligence gathering and monitoring of Japanese scientists. On September 13, 1945, the postal system in Japan was taken over by the CCD, thus enabling SCAP to scrutinize what was being communicated throughout the country.[56]

Internal mail provided a major source of information. Rallies and large-scale meetings involving Marxist scientists were undoubtedly the target of security surveillance, as was Minka itself.[57] Although there is some conflict over the exact number of members and branches, one intercepted letter dated mid-1949 provides the following information. The stated objectives of Minka were (1) to protect and maintain freedom of research; (2) the popularization of science; and (3) to construct a grassroots-level type of democratic science, which would relate to everyday life. Membership tended to be centered in Tokyo (2500 members in 50 branches) with 8000 other members organized in 70 branches. This network of scholars belonged to different sections within the Association: Economics, Philosophy, History, Jurispudence, Agriculture, Art, Philology, Natural Science, Anthropology, Psychology, and Pedagogy.[58]

Censorship control was relaxed in the second half of 1947, but censorship of what was deemed "ultra-left," communist material continued even after this period. By the end of 1947, a shift of emphasis from prepublication censorship of publishing to postpublication censorship had occurred, enabling CCD to expand its activities of monitoring subversive activities, intelligence gathering, and analysis. Surveillance included examination of mail, telegrams, and telephone calls.[59]

Around 1947–1948, with the onslaught of the Cold War, SCAP became less tolerant of the calls for social "revolution" and adopted what is sometimes called the "reverse course" in its policies toward workers' organizations and leftist groups. On June 25, 1948, the Japanese English language newspaper *Nippon Times* reported how Woo Kya-tang, the executive editor of the American-owned Shanghai Evening Post and Mercury, viewed that a communist Japan was the most urgent threat to East Asia.

He said "an economically bankrupt communized Japan under remote control from the Kremlin today is the most urgent threat to the tranquility of the entire Pacific area. . . . There cannot be any revival of Japan's economic military setups to an extent approaching a real threat either to China or to the United States within 25 to 50 years," he said. "Far more immediate than the mental picture of a new militaristic Japan in some far distant future," the writer stated, "is the constant increase in the influence of Japan's communistic organizations. The lack of food, the low economic levels of the Japanese masses and general economic maladjustment all contribute toward a rapid rise of communization of Japan if the Red spark is ever allowed to spread" he said.[60]

That same month, the Japanese Government denied that it had started an anticommunist drive. There had been claims that the attorney-general's Office had begun investigations into five communist organizations with the view of dissolving them.[61] It is no coincidence that Sakata's writings and correspondence were monitored by SCAP. An example of a letter intercepted by the CCD was a communication from Sakata to the Physical Society of Japan, postmarked August 18, 1948. The letter listed topics on which Nagoya University physicists would speak at a meeting on nuclear theory sponsored by the Society. The letter was passed.[62]

While Sakata does not appear to have been a member of the Communist Party, his public endorsement of its science policy was a matter of concern to SCAP. The following pronouncement was duly filed by ESS:

I hope that as many democratic forces as possible will be elected in the coming general election, because the Ministry of Education draft of the University Law, which was mapped out by the conservative reactionary forces and introduced by the Ashida Cabinet, hampers the free development of Japan's academic circles and has a tendency to colonize the nation's education. I see fascism in it. In order to cope with this fascism we must unify all progressive forces.

From this point of view the current trend of social Communism coup which is seen throughout the country is indeed a matter for congratulations. Labor's political party should work for labor. The Communist party's policies in regard to science and education are what we scientists are demanding.[63]

Earlier that year, it was reported in the Japanese press about how President Truman had denounced communism as a "false philosophy" in his presidential inauguration speech. It was reported in *Nippon Times*, January 22, 1949, that

With television cameras recording an inauguration for the first time, the great throng volleyed applause as Mr. Truman pitted the blessings of democracy against the evils of Red Marxism. "That regime adheres to a false philosophy which purports to offer freedom, security and greater opportunity to mankind," he asserted. "Misled by this philosophy, many people have sacrificed their liberty only to learn to their sorrow that deceit and mockery, poverty and tyranny, are their reward. That false philosophy is communism."[64]

For such reasons, Taketani's outspokenness on social issues attracted the continued attention of SCAP. Sakata, Taketani, and many other Japanese intellectuals championed the Marxist perspective. Soviet Marxism served to reinforce the view of the importance of scientific method. Dialectical materialism could be used to interpret both nature and human affairs. As J.D. Bernal argued,

> It is to Marxism that we owe the consciousness of the hitherto unanalysed driving force of scientific advance, and it will be through the practical achievements of Marxism that this consciousness can become embodied in the organization of science for the benefit of humanity.[65]

Sakata and Taketani, even in their accounts of the development of the Nobel Prize–winning meson theory, relate how Engels' *Dialectics of Nature* helped to order their research in terms of three stages of cognition of nature: the phenomenological stage in which the phenomenon is described; the substantialistic stage in which its structure is investigated; and the essentialistic stage in which interactions and laws of motion of the object are clarified.[66] This represented an attempt to demonstrate how Marxism would show the way. The direct application of dialectical laws through the three-stage theory was said to have been important in the development of meson theory. Using those laws in social practice would be an even greater challenge. For Sakata and Taketani, Marxism and science were meta-narratives that provided large-scale, theoretical interpretations of the world, which supposedly had universal application.

In the Japan Science Council (JSC) elections held in December 1948, Sakata was elected at the relatively young age of 38. He would be continually reelected a further seven times, devoting 22 years of his life to JSC activities.[67] He and Gorō Hani lobbied for research autonomy through the Committee for Freedom of Thought and Learning (Gakumon Shisō no Jiyū Hoshō Iinkai), which they helped to establish. The Committee conducted a nationwide survey of scientists. Sakata was dismayed to find that many considered the Pacific War as the time of greatest freedom in research. He interpreted this finding as meaning that abundant research funds were available then, but he berated them for their poor understanding of the true meaning of research autonomy. He was fearful that the large sums of money being made available for atomic energy would again tempt scientists.[68]

Sakata was all too aware that the fervor for reconstruction after the war might mean the suppression of freedom in scientific research in favor of specific applications of science. He was a member of the JSC Committee for Freedom of Thought and Learning for the first two terms of the Council, from 1949 till the beginning of 1954.[69] It is significant that at the time of the Red Purge and emergence of the Cold War, Sakata would push for professional autonomy.

Sakata, like Yukawa, received his share of accolades. On January 16, 1949, he was awarded the Asahi Prize for his two-meson theory and shortly thereafter, on January 20, became a member of the Science Council, along with Taketani (figure 4.1). In 1950, Sakata was awarded the Imperial Award of the Japan Academy for the two-meson theory and was reelected to the Science Council the following year.[70] The first term of the Science Council lasted two years, but succeeding terms were of three years duration. In the second term of the Science Council from January 1951 to January 1954, Kameyama and Wagatsuma retained their posts and were joined by Seiji Kaya as vice-president. Yō Okada became the chairman of Division Four and the vice-chairman was Yoshio Fujioka. The secretaries were Yūsuke Hagiwara and Shōichi Sakata. District area organizers were appointed and for the central Chūbu area, the physicist Kanetaka Ariyama of Nagoya University was elected.

Also elected to the Science Council, Taketani was the subject of an interview in *Minshū hyōron (People's Review)* in June 1949. Taketani criticized the accuracy of reporting in all the major Tokyo papers. He pointed out how a bloody skirmish between police and Hitachi mineworkers had been ignored by the press. Commentary on unions tended to be on their internal conflict and editors had no opinion of their own on the civil war in China. Taketani suggested that a weekly publication be produced, which would systematically act as a media watchdog,

> reviewing on a nation-wide scale, all newspaper articles on peace movements sponsored by intellectuals, and in warning the publishers against any possible mistake they may make in the handling of news concerning such movements.[71]

Taketani would remain a watchdog for the rest of his life, especially when it came to the problem of harnessing atoms for peace. Like other physicists, he stressed that the value of physics lay in the physicist's

Figure 4.1 Shōichi Sakata and Mituo Taketani at the Japan Academy, Ueno, Tokyo, where a General Meeting of the Science Council of Japan was held on April 28, 1951.

Photograph: Ken Domon. Courtesy, late Mrs. Nobuko Sakata and Prof. Fumihiko Sakata.

moral character, an ideal type of person.[72] Taketani was fashioning himself as a public man.

IV   ATOMS FOR PEACE

During the Occupation, writing on the effects of the atomic bombs on Hiroshima and Nagasaki was heavily censored.[73] Nevertheless, Taketani wrote on nuclear war and suggested that the threat of mass annihilation was serving as a brake to further global conflict. He was also the first academic to push for the peaceful use of nuclear power in Japan. In 1950, Taketani published a volume on the topic through the newspaper publishing company Mainichi Shinbun. This book entitled *Genshiryoku (Atomic Energy)* explored the possibility of the use of the atom as a peaceful way of obtaining electrical power.

The end of the Occupation in 1952 provided many opportunities for Taketani to write on the subject of the atomic bomb and atomic energy. Particularly notable was his contribution to the special issue on the bomb in *Kaizō* (October 1952) in which he interviewed the director of the Atomic Bomb Casualty Commission (ABCC) in Hiroshima. A discussion of "The Smyth Report"[74] appeared in *Asahi hyōron* and he edited *Genshiryoku (Atomic Energy)* (Tokyo: Mainichi Shinbun Sha, 1950) with assistance from Seitarō Nakamura.[75] Soon after, in an article in the periodical *Kaizō* in November 1952, Taketani wrote that although he felt that Japan should pursue atomic energy, there were two problems: (1) insufficient domestic supply of uranium, and (2) the danger of ending up as a "subcontractor" for the U.S. atomic bomb program. Taketani felt that as the world's first victim country of the atomic bomb, the Japanese had a right to conduct research on the peaceful uses of atomic energy, with the assistance of overseas countries. He felt that such countries should unconditionally supply Japan with the necessary quantities of uranium, and that atomic energy research should not be of a secretive nature.[76] By his writings, Taketani was instrumental in educating the public about the facts and implications of atomic energy.

In February 1952, Taketani called to the Japanese to build their own nuclear reactors in order to break the stranglehold on such know-how held by nations possessing nuclear weapons.[77] Although research into atomic energy was prohibited during the Occupation, Sakata and Taketani, nevertheless, felt that it was important to discuss means of ensuring that such research, when allowed, would remain autonomous. There were fears that the prohibition would remain in place even after a Peace Treaty was signed, but this was not so. The Peace Treaty came into effect in 1952, and the prohibition was removed.

Sakata had hoped that prior to commencing research into atomic energy, a powerful research organization like France's Centre National de la Recherche Scientifique (CNRS) might be established for fundamental research and that this might provide a framework for the peaceful uses of atomic energy. Sakata drew inspiration from the communist physicist, Frédéric Joliot who was director of CNRS from August 1944. Joliot was the son-in-law of Marie Curie and joint-recipient of the 1935 Nobel Prize for physics with Irène Joliot-Curie. Like Sakata, he had hopes of using science for the welfare of the people and entertained hopes that a national science agency could be run by scientists. Joliot and his colleagues drafted the ordinance for a Commissariat à l'Energie Atomique (CEA), which Charles De Gaulle promulgated on October 18, 1945. The CEA was unusual in that it was a largely civilian nuclear agency, which was ruled by an Atomic Energy Commission. Joliot was high commissioner and took charge of CEA's science and technology programs, as well as helping to found the World Federation of Scientific Workers in July 1946.[78]

Sakata feared the establishment of an Atomic Energy Commission and a Science and Technology Agency controlled by the Japanese government, which would emphasize applied science over basic, and lead to the remobilization of science for military purposes.[79] The Committee for Freedom of Thought and Learning was an early forum for discussion about such fears. In 1952, debate extended to all of the Science Council. In an interview, which was published in a November issue of the magazine *Kaizō*, Taketani confronted the director of the ABCC regarding its investigations of the radiation effects on the people in Hiroshima and Nagasaki. He suggested that the real role of the ABCC was to study atomic bomb countermeasures in case of future war.[80] Many JSC members opposed military cooperation and were reluctant to embrace atomic energy. Events soon overtook them.

In 1954, a budget allocation for a nuclear reactor was approved by the Diet. The Special Committee for Nuclear Research (SCNR) proposed three principles for atomic energy.[81] Taketani berated the 1954 Overseas Fact-finding Mission for Atomic Energy led by Yoshio Fujioka for ignoring the views of the JSC SCNR. Fujioka had joined the government's Atomic Energy Commission as a representative of physicists but had alienated the very physicists whom he supposedly represented.[82] The question of how Japan should approach atomic energy divided physicists.

Taketani first proposed the three principles for the peaceful use of atomic energy, calling for public disclosure of atomic research, democratic management of research, and research autonomy. It was hoped that they would prevent atomic research from being used for military purposes. Science, in Taketani's mind, needed to be conducted in an open manner and made accessible to most people. SCNR agreed with this and passed on the proposal for approval to the Thirty-Ninth Committee, which dealt with atomic energy issues. It was later ratified in a statement issued by the seventeenth general assembly of the Science Council of Japan. Through this process of consultation, all levels of the Science Council became acquainted with the principles and involved in their advocacy. The initial authorship of the principles is, for this reason, little known. It is difficult to trace the instigators of policies formulated by the Science Council because of this collective approach to policy-making.[83] Atomic energy became the focus of discussions in the Science Council. Sakata's advocacy of the importance of the "three principles" ensured that public debate over atomic energy continued.

Sakata and Taketani's actions had a major impact on the formulation of the Basic Atomic Energy Law passed by the Diet in 1955, which forbade use of atomic energy for military purposes and which also incorporated the three JSC principles. The commercialization of atomic energy has meant that the management of research has not always been democratic and autonomous, and open access has not been guaranteed, but the militarization of atomic energy has been strongly resisted.[84] The "Three Non-Nuclear Principles," declared by Prime Minister Eisaku Satō in 1967, amounted to a restatement of the three principles of JSC and the Basic Atomic Energy Law. Japan would neither possess nor produce nuclear weapons, and would not allow nuclear warheads in its territory.[85]

In 1956, the Japan Atomic Energy Commission (JAEC) was established but, to the physicists' dismay, its enthusiastic pursuit of atomic energy reflected a nation that had yet to come to terms with its own wartime guilt and had conveniently forgotten the devastation that atomic energy could wreak.[86] Yukawa was a member of the Commission but resigned within a year. This reflected the difference in approach between academia and the industrial sector. These differences were also manifested in the debate surrounding the safety of the Calder Hall reactor, which was introduced into Japan in 1958. Although Sakata was a member of the Reactor Safety Investigation Committee (Genshiro Anzen Shinsa Senmonbukai) of the JAEC, he too would resign in protest about the lack of consultation.[87]

Sakata was continually reelected to the Science Council until the time of his death. He was chairman of the Committee for Atomic Energy Problems and the SCNR for long periods. The SCNR had been the National Committee for Nuclear Research, chaired by

Yoshio Nishina, which prepared reports for the Occupation Forces on
Japanese wartime nuclear research. The SCNR comprised of specialists from theoretical and experimental nuclear physics, and cosmic-ray
physics. Until 1956, the SCNR had been chaired by Sin-itirō
Tomonaga, after which time Shōichi Sakata took over the chairmanship, assisted by the vice-chairmen Seishi Kikuchi (director, Institute
for Nuclear Study) and Yuzuru Watase (Osaka City University).
The SCNR was particularly concerned about (1) the creation of
research institutes for joint-use; and (2) problems relating to the
development of nuclear power. It was supportive of the establishment
of the Research Institute for Fundamental Physics (RIFP), which
opened in 1953 at Kyoto University.[88] SCNR was also instrumental in
lobbying for the creation of a joint-use nuclear research facility
equipped with accelerators. In 1953, the general assembly supported
such a plan and their efforts were realized with the establishment of
the Institute for Nuclear Study at the University of Tokyo. The
Cosmic Ray Observatory at Mt. Norikura was also established for
joint use.[89] In April 1954, the Thirty-Ninth Committee (for atomic
energy problems) was dissolved and was succeeded by a new committee, the Committee on Problems in Atomic Energy. This committee
was chaired by Shōichi Sakata, and Hirokatsu Ogura and Yōichi
Fukushima were the secretaries.[90] The Special Committee for Atomic
Energy chaired by Kōji Fushimi (then at Osaka University) was
established in early 1956.

In 1955, the Russell–Einstein Manifesto was released. In April
1956, Sakata attended a special assembly of the World Council of
Peace in Stockholm, reporting on Japan's three atomic energy
principles and the movement to ban atomic and hydrogen bombs. In
his talk before the assembly, Sakata felt that Japan's recent agreement
to receive American financial assistance in the construction of nuclear
reactors might render Japanese atomic energy research classified or
place it under the control of the United States in future.[91] In 1957,
the Pugwash Conference was established, serving to place the debate
on atomic energy even more within the context of international issues
such as arms control rather than just domestic energy concerns. Sakata
attended the Third Pugwash Conference on Science and World Affairs
held in September 1958 in Kitzbuhel and Vienna. The meeting
reflected the concerns of scientists who attended: "The Dangers of
the Atomic Age and What Scientists Can Do about Them."[92]

The year 1957 was one of special concern to the Science Council.
On November 27, State Minister Matsutarō Shōriki, director of the
Science and Technology Agency, suddenly announced that he wanted

to establish a committee for the promotion of science and technology not unlike the JAEC. By 1958, a bill had been passed by the House of Representatives, which established a Science and Technology Council (Kagaku Gijutsu Kaigi), headed by the prime minister and including four scholars and one representative from each of the Ministries of Finance and Education, Science and Technology Agency, and Economic Planning Agency. This was done without consultation with the Science Council.[93]

In 1960, Sakata was chosen to chair the Special Committee for Atomic Energy, into which the former Committee for Atomic Energy Problems was absorbed. Sakata, Yukawa, and Tomonaga proposed organizing a Japanese version of the Pugwash Conference. The first was held on May 7–9, 1962 and called the Kagakusha Kyōto Kaigi (Kyoto Conference of Scientists).[94] A second conference was held the same time the following year at Takehara, near Hiroshima, focusing on the themes of "the World Situation after the Cuban Crisis," "Japan in Asia," and the "Social Responsibility of Scientists." A statement issued by the Conference asked that the People's Republic of China be admitted into the United Nations.[95] This stance of encouraging interaction with China was reinforced in 1964 when Sakata and other Japanese physicists attended the Peking Symposium. Sakata was the Japan delegation leader.[96]

In April 1965, Sakata, Tomonaga, and Yukawa sent a telegram to the fourteenth Pugwash Conference urging the withdrawal of U.S. troops from Vietnam and requesting cessation of chemical warfare. In 1966, Sakata became the dean of the Faculty of Science at Nagoya University. In addition to such administrative duties, he continued to be concerned with social issues and contributed to the Third Kyoto Conference of Scientists, which was held in Tokyo, June 30–July 2, 1966. The Conference addressed the questions of "balance of power" and the potential for war in space.[97] In April 1967, Bertrand Russell wrote twice to Sakata on behalf of the Bertrand Russell Peace Foundation and the International War Crimes Tribunal. Both letters concerned opposing American aggression in Vietnam. The misuse of scientific resources in the Vietnam War for destructive purposes reinforced the view that the natural sciences and capitalism were not compatible.[98] Despite his involvement in such political activities, Sakata was still held in high esteem by physicists for his physics, participating in the Solvay Conference in Brussels in September 1967. The following year, a book coauthored by Sakata, Tomonaga, and Yukawa, entitled *Kaku jidai o koeru (Beyond the Nuclear Age)* (Tokyo: Iwanami Shoten, 1968), was published.

On reflecting over two decades of the Science Council shortly before his death, Sakata remarked on how the prewar National Research Council had consisted of "goyōgakusha" ("government-patronized scholars" or "unprincipled scholars") who toed the government line. The "logic of government" dominated the "logic of science." In the 20 years since then, the Science Council had been active, but since the time of the Shigeru Yoshida cabinet, moves had been underway to bypass the Science Council. Rather than applying the logic of science to government policy-making, government was smothering science with its logic, such as that of the pursuit of short-term profit.[99] Taketani and Yōichi Fujimoto suggested that the government had used the excuse that they were advancing "big science" to build an administrative structure for science and technology, which ignored the Science Council and was bereft of scientists.[100]

With failing health, Sakata resigned from his long-running chairmanships of SCNR and the Special Committee for Atomic Energy (both in one form or another since 1957). He passed away on October 16, 1970.[101] His death marked the end of a period of the physicist as a general expert on things scientific. The increased number of specialists made "expertise" more accessible and along with it any political influence that expert authority wielded.[102]

V   THE ELEMENTARY PARTICLE THEORY GROUP

It is important to understand that Sakata and Taketani's activities and concerns reflect those of a larger collection of physicists, the Elementary Particle Theory Group. The origins of the group can be traced to symposia on meson theory, which were held before and during World War II, at annual meetings of the Physico-Mathematical Society of Japan, or at Riken lecture meetings. Those who participated in such meetings numbered around 20 and tended to belong to Riken or to the imperial universities of Tokyo, Kyoto, Osaka, and Nagoya.[103] These "meson meetings" provided gatherings for open discussion on particle physics for physicists who would be leading members of the group after World War II: Hideki Yukawa, Sin-itirō Tomonaga, Shōichi Sakata, and Mituo Taketani.

Scientists felt a keen sense of social responsibility, and through organizations such as the Science Council (established in 1949) and the Elementary Particle Theory Group, sought to influence science policy and bring about the "modernization" of the research system. In order to improve access to research literature and to improve liaison within the Group, at the end of 1949, a network consisting of nine branches was set up throughout Japan. These branches were located at the Universities of Hokkaido, Tohoku, Tokyo Bunrika, Tokyo, Nagoya, Kyoto, Osaka, Hiroshima, and Kyushu.[104] Through membership of the Group, less well-connected physicists in the regional universities could enhance their status, be empowered, and gain considerable prestige through association with high-profile physicists such as Yukawa, Tomonaga, Sakata, and Taketani. By 1950, the membership of the Group had grown to well over 100, with some estimates as high as double that figure.[105] This was helped along by the awarding of the 1949 Nobel Prize for Physics to Hideki Yukawa. The subsequent euphoria, which resulted from the award, has been labeled the "Yukawa Effect."[106] In addition to the establishment of various scholarships, a building in honor of Yukawa ("Yukawa Hall") was constructed on the campus of Kyoto University.[107] The Hall was opened on July 20, 1952. One of the results of such attention was that the group became something of a favorite of the media.[108] Japan's first Nobel Prize had a real impact on Japanese attitudes to science. The National Diet passed two resolutions recommending an increase in government funding of R&D as a result. This was in the House of Councillors on November 28, 1949 and the House of Representatives on March 11, 1950.

The Elementary Particle Theory Group combined the comaraderie and collaborative approach to research, which characterized the wartime meson theory meetings, and the democratic spirit of the postwar period. Its mission was to encourage joint research nation-wide as well as to provide a forum for issues relating to research. The Group, along with two other groups representing nuclear physics, would air their views through the SCNR, which came into being in 1952, succeeding the Liaison Committee for Nuclear Research. From 1957 to shortly before his death, Sakata headed this Committee. It not only acted as a planning body for nuclear physics, but also served as a body that would engage in the atomic energy debate.[109]

The RIFP at Kyoto University was a product of the need of groups, such as the Elementary Particle Theory Group, for cooperative research institutions run in a democratic manner.[110] In 1953, the RIFP was established in Yukawa Hall, headed by Yukawa who became the first director and with Sakata as a member of the Management Committee. Although the research institute was attached to Kyoto University, its management differed from that of other university-affiliated institutes. It was run with considerable input from researchers in the Elementary Particle Theory Group and the Solid

State Physics Group.[111] RIFP served as a gathering point for the Elementary Particle Theory Group, and the location of study meetings, and dissemination point for literature and information from abroad. The English language journal entitled *Progress of Theoretical Physics* had been launched in Japan in 1946 under the editorship of Hideki Yukawa. This journal sought to provide a wider audience for the works of Japanese theoretical physicists. In addition to *Progress of Theoretical Physics* and its *Supplement*, a mimeographed circular entitled *Soryūshiron kenkyū* (*Elementary Particle Theory Research*) is also published by the research institute. *Soryūshiron kenkyū* consists of research papers in the Japanese language, and also serves as a means of communication among members of the Group.[112]

In 1950, the Elementary Particle Theory Group began to hold regional conferences under sub-group named REKS (Research Group of Elementary Particles in Kansai) and REKT (the Kanto equivalent, which met in Tokyo). As there were also two meetings of the Physical Society of Japan per year, physicists were attending conferences every three or four months, in addition to attending REKS and REKT research seminars.[113] The Elementary Particle Theory Group attempted to further improve its organization and in 1952, set about implementing the following policies. It was decided that seminars would be held on a regular basis, thereby strengthening contact between Group members. Academic vacancies at research institutions would, as far as possible, be advertised and be open to applicants. These policies would be facilitated by a rotating secretariat, the responsibility for liaison thus being shared by a number of universities. Part of research funds would also be pooled to enable scholars to attend seminars conducted in other parts of Japan. This was labeled the "Mushashugyō" ("Warrior Training") system—a modern-day ritual aimed at nurturing young physicists who were almost always men.[114]

Despite the democratic aspirations of the Elementary Particle Theory Group, a definite hierarchy has existed, based upon generational lines. The members of this hierarchy have been affectionately labeled "daibosu" ("big bosses"), "chūbosu" ("middle bosses"), and "kobosu" ("little bosses"). Hideki Yukawa, Sin-itirō Tomonaga, Shōichi Sakata, and Mituo Taketani were the first generation of bosses. Unlike his three colleagues, Taketani spent many years on the academic periphery, writing on social issues for a variety of magazines and journals. Such writing not only provided him with an income but helped him to fulfill his sense of duty to educate Japanese citizens about key issues relating to science and technology, to help them understand the new world that they were now living in. It was only from 1953 that Taketani was employed on a full-time basis as an academic at Rikkyo University in Tokyo in theoretical physics.[115] This famous four were followed by a group of physicists who graduated from university around the time of World War II. This group, called the middle bosses, included scientists such as Seitarō Nakamura, Takeshi Inoue, and Nobuyuki Fukuda. They also included a group of physicists who completed university studies at the end of the war or soon after: Satio Hayakawa, Yōichi Fujimoto, Hiroomi Umezawa, and Eiji Yamada. The little bosses is the term used to describe a postwar generation of physicists who have followed on from these physicists.[116]

As Hirosige has written, what is known as the Elementary Particle Theory Group is an amalgam of three things: a research group for the purposes of the Science Council of Japan, a sectional meeting of the Physical Society of Japan, and an open-ended, spontaneous gathering of researchers.[117] By the end of the 1950s, membership of the group exceeded 300, enough to warrant thinking of it in terms of a learned society.[118] The group has not been a clearly defined body like the Democratic Scientists' Association or the Physical Society of Japan. It has, however, constituted a formal group for the funding purposes of the grants-in-aid of the Ministry of Education.

After the establishment of the RIFP in 1953, other joint-use research institutes were created: the Institute for Nuclear Study at the University of Tokyo (INS) in 1955; the National Cosmic Ray Institute; the Tokyo University Institute for Solid State Physics (1957); Osaka University Protein Research Institute (1958); Nagoya University Institute of Plasma Physics (1961); and the Kyoto University Research Institute for Mathematical Analysis (1963).[119] The Elementary Particle Theory Group has been a major user of the facilities of the Research Institute for Fundamental Physics (Kyoto University) and the Institute for Nuclear Study (Tokyo University).

The Group has also been outspoken on issues such as the renewal of the U.S.–Japan Security Treaty and nuclear power.[120] Its membership has consisted of scientists working in areas such as quantum field theory, particle theory, nuclear theory, and cosmic-ray physics. Communication within the group has been facilitated by the mimeographed circular *Soryūshiron kenkyū*, which was first published in 1948.[121] Other groups for physicists were created, using the Elementary Particle Theory Group as a model: "Uchūsen Kenkyūsha Kaigi" ("Cosmic Ray Researchers Council" or "CRC"), "Genshikaku Danwakai" ("Nuclear Physics Forum" for experimental physicists), and the "Busseiron Guruupu" ("Solid State Physics Group").

The Elementary Particle Theory Group has been one of the most active groups in the movement against atomic and hydrogen weapons.[122] In May 1957, it supported the "Gottingen Statement" which was issued on April 12, 1957 by 18 German physicists including Werner Heisenberg. The Group has supported the activities of the Pugwash Conference and, along with the Science Council, appealed for the suspension of nuclear testing. In 1957, a delegation of 20 Japanese physicists led by Sin-itirō Tomonaga visited China to encourage dialogue between the scientists of both countries.[123] Sakata had visited the Soviet Union and People's Republic of China the previous year. These trips were made at a time when the two countries enjoyed no formal diplomatic relations with Japan. Sakata's interest in China and the Soviet Union perhaps lay in their embodiment of the idea that science could not truly advance if a capitalist system was maintained.[124] What future has science then in what has now become one of the richest nations in the world?

The Elementary Particle Theory Group can be cited as a particularly significant group for three reasons, apart from the group's contributions to physics: (1) the contribution of its members in the democratization of Japanese academic research; (2) their awareness of the social responsibilities of scientists; and (3) their use of dialectical materialism to guide their research.[125] To what extent the latter has been useful is something open to considerable debate.

CONCLUSION

Sakata's sense of public vocation can be traced to his samurai origins. Like his ancestors before him, Sakata displayed a sense of public responsibility and commitment to "national" interest and welfare, as interpreted by him, progressive scientists, and other Marxist intellectuals. Taketani lacked the privileged background of Sakata, but, nevertheless, adopted this feeling of public vocation like other intellectuals in postwar Japan. He provides an excellent example of how the samurai tradition of public service was appropriated and perpetuated by the intellectual class,[126] with Marxism as the ideology informing their discourse.

But the Japanese Marxist intellectual has been accused by Edwin Reischauer of being more "intellectual aristocrat and pure theoretician rather than a practical leader, standing haughtily and a little tragically aloof from society." Reischauer concedes that after World War II, Marxist scholars showed greater political consciousness, but a continuing emphasis on a theoretical approach to the problems facing

Japan meant that it "has left them with almost as little practical influence as before."[127] Sakata and Taketani tried to transcend this stereotype of a Marxist; the former by reshaping Japan through its research system and research policies, and the latter through the awareness of social issues.

Taketani, in his many writings over the years, has defined what public issues physicists can address with their special knowledge. Physics was seen as being more fundamental than the other sciences. This enabled physicists to transcend the specificity of their work and deal with more universal issues. The threat posed by nuclear weapons provided Taketani, Sakata, and their colleagues with universal problems that needed to be urgently addressed.[128] Taketani's writings helped form the basic parameters of public discussion on atomic energy. Sakata and Taketani stand out as two physicists who broadened the concept of "public interest" to include the social needs and welfare of the Japanese population rather than some abstract national interest or specific military concern. Their relatively egalitarian approach to science differs from that of Yukawa and Tomonaga. What has been written of J.D. Bernal and his colleagues equally applies to Sakata and Taketani:

> What he and others of his generation had changed was the definition of what constituted the public good or national interest that science should serve.[129]

Physicists co-opted discourses from the public domain to suit their purposes. But for Sakata and Taketani democracy was more than a catchword; it was a way of changing their world. The demand for reform and modernization in Japan, and the hope that the reorganization of science and technology would help achieve that, was an exciting chance for their vision to be realized. The Elementary Particle Theory Group facilitated this.[130] Many members felt that their social and research activities were intermeshed. Science was viewed as essentially value-free. "Good science" might sometimes be used for "bad" purposes, but it would be one of the responsibilities of fellow physicists to ensure that it was not. Their social role was to thus promote science (especially basic research) as an exemplary form of knowledge, which could only progress if scientists (especially physicists) were given control and were able to protect science from those seeking only to profit from it. This interpretation of the social responsibility of science was an essentially political one, which varied depending upon the actor. Such beliefs guided Sakata's actions in the

Science Council, the major entry point for scientists into policy-making. The history of the Science Council is a story of conflict over authority in policy-making, the power of the state versus the desire for professional autonomy. It also is a story of differing visions for science: a tool by which to bring about rapid change or part of an academic quest for knowledge with potentially useful results.[131]

Despite Sakata's lobbying through the Science Council and Taketani's extensive writings, both remained relative "outsiders" in government policy-making, seeking to influence the establishment from outside. Their activities reflect how

> The intellectuals spread discontent and give systematic form to criticism and alternative possibilities; they take a leading part in the work of transformation and modernization; and once the new institutions are created, they often take a leading part in their operation.[132]

Sakata and Taketani encouraged scientists to participate in public affairs. Partly in order to counteract the left-leaning tendencies of such scientists who viewed American science and its links with the military with suspicion, a U.S.–Japan Cooperative Science Program was established in 1961.[133] By taking into consideration this sort of impact and the role of intellectuals in policy-making, a fuller, more pluralistic image of power and influence in Japan emerges. The degree of this influence is difficult to measure, but it is clear that it is of considerable importance. It is often even more so when the intellectual is able to operate from within the establishment. The following chapters provide examples of where this has been the case.

CHAPTER 5

THE POLITICS OF PURE
SCIENCE: YUKAWA AND
TOMONAGA

In 1954, the U.S. Information Service (USIS) produced the film "The Yukawa Story," which was based on the life of the Nobel Prize–winning physicist Hideki Yukawa who had recently worked in the United States and returned to Japan to great acclaim. Narrated by one of his two sons, Takaaki, the film documented the life and work of Yukawa. Takaaki relates to the audience how he faced an

Initial dilemma in reconciling his father's world with his mother's traditional one. He finally realizes that inductive reasoning and emotional appreciation of cultural heritages are synthesized in his parents' lives.[1]

The film was part of an attempt by the U.S. government to ensure the continued orientation of Japan toward the West. The pro-U.S. "Yukawa Story" helped to balance the influence of leftist intellectuals such as Sakata and Taketani. By 1955, the USIS had funded six feature films, under contract with Japanese production companies, which were unattributed to the USIS.[2]

We saw in the previous chapter how the "big bosses" of physics, Sakata and Taketani, strived to shape science policy through "laboratory democracy," the Science Council, and extensive publications, using their expertise in physics as their authority. In the eyes of the public, however, Sakata and Taketani remained very much cast as social spokesmen, their ideology sometimes dominating their science.

Hideki Yukawa and Sin-itirō Tomonaga also were big bosses, but they influenced the direction of research in a different way from

Sakata's and Taketani's. They achieved their reputation foremost by their work in physics, as recognized by various prizes and ultimately, for both, the Nobel Prize. This lent them an aura of credibility, which Sakata and Taketani had not quite achieved. Yukawa and Tomonaga, as internationally respected physicists, provided stability to a Japan caught up in the confusion of rapid economic growth. They were seen as holding the promise of bringing order into national life. The rationale was that scholars who have been "prized" internationally deserve to be listened to on a variety of matters not always connected with their area of specialization. These "experts" were public men in that they were perceived as possessing general qualities showing that they were men of ability. Given the linkages between science, democracy, and Japan's social and economic reconstruction, which Sakata and Taketani had helped to establish, it is little wonder that the Japanese embraced Sin-itirō Tomonaga and Hideki Yukawa as experts to whom they could defer. For these two physicists (and especially the latter), science was an elite activity. That this contradicted Sakata's and Taketani's message of the need for representation, participation, and equality was not seen as being problematic.

The first two sections of this chapter provide short biographies of Japan's two Nobel-Prize physicists, Yukawa and Tomonaga, which help us to understand why their life-stories and achievements might be appealing to both sides of politics. Family background and education help to account for differences in research styles as discussed in the third section. The Nobel Prize reinforced their authority and served to distinguish them from Sakata and Taketani. The fourth section focuses on how this assisted them in their activities as "peacemakers" while gaining an international voice by championing the cause of nuclear disarmament.

I   HIDEKI YUKAWA

For Hideki Yukawa, cultural traditions were an important part of growing up. In his biography, *Tabibito*, and philosophical writings on science, he suggests how such traditions were formative for his science as well.[3] These writings, combined with his status as Japan's first Nobel laureate, and his long directorship of the Research Institute for Fundamental Physics earned him a special place amongst physicists.

Family, education, and marriage shaped Yukawa's identity. He was born in Tokyo on January 23, 1907, the third son of Takuji Ogawa. Takuji was a staff member of the government Geological Survey Bureau, part of the Ministry of Agriculture and Commerce.[4] In 1908, shortly after Hideki was born, Takuji took up an appointment as professor of geography in the Faculty of Arts at Kyoto University; so the family moved to Kyoto. Hideki's grandfather, Nanmei Asai, was a former Confucian scholar for the Tanabe feudal clan. Nanmei had taught the Chinese classics at the clan school; hence, it is not surprising that Hideki and his brothers would also receive such lessons from their grandfather. Hideki's maternal grandfather, Komakitsu Ogawa, had served as a samurai at the Tokugawa castle in Wakayama. Komakitsu, too, possessed a good understanding of the Chinese classics. His daughter Koyuki married Takuji Asai, who was subsequently adopted into the Ogawa family. Takuji changed his name to Ogawa to continue the Ogawa family line.[5]

Intellectual authority required formal qualifications. Hideki began obtaining his at Kyōgoku Elementary School, Kyoto, in 1913, the second year of the Taishō period (1912–1926), going on to Kyoto Prefectural First Middle School in 1919. He entered the Third High School in 1923 and gained admission to Kyoto Imperial University in 1926. Shortly after entering university, Yukawa attended a special lecture given by Hantarō Nagaoka (at that time based at Tokyo Imperial University) on the past and present state of physics. Yukawa was deeply impressed.[6] He graduated in 1929 and that year stayed on as an unpaid associate in the lab of Professor Kajurō Tamaki in the Faculty of Science, pursuing theoretical physics. The Tamaki lab did not formally take on postgraduate students, but Yukawa and his colleague Tomonaga found the research environment congenial enough and the job market bad enough to warrant staying on.[7]

Hideki Ogawa married Sumiko (Sumi) Yukawa on April 23, 1932, and was adopted into her family,[8] taking on the family name of Yukawa and moving to nearby Osaka to live. Sumi was the eldest daughter of the owner of the well-known Yukawa Gastrointestinal Hospital, Genyō Yukawa and his wife, Michi. Genyō too had been adopted, his former name being Josaburō Sakabe, a descendant of a prominent samurai family.[9] Hideki was appointed lecturer that year at Kyoto's Faculty of Science, teaching quantum mechanics and carrying out theoretical research in nuclear physics and cosmic rays. The second year students Shōichi Sakata and Minoru Kobayashi attended Yukawa's first lectures. Mituo Taketani would take the quantum mechanics course the following year.[10]

At a meeting of the Physico-Mathematical Society of Japan held at Tōhoku University in April 1933, Yukawa met Hidetsugu Yagi who was about to leave that university and head a physics department at the new Osaka Imperial University. That year, Yukawa became a lecturer at

Osaka Imperial University, continuing to teach also at Kyoto until March 1934.[11] At the time of his employment at Osaka Imperial University, Hantarō Nagaoka was university president. The pioneering experimental nuclear physicist Seishi Kikuchi also joined the physics department.[12]

By the 1930s, three main types of forces were known to be working within the atom. Electromagnetic force was associated with the interaction of charged particles. "Weak" force was involved in beta decay, and "strong" force was connected with the binding force of the nucleus. Yukawa thought that the nuclear force could be reduced to the exchange of a new particle between the proton and neutron—the meson. The interaction would be described by a new field of force, just as the interaction between charged particles was described by the electromagnetic field.[13] Yukawa developed his ideas regarding the meson while at Osaka, and on November 17, 1934, delivered a paper at a meeting of the Physico-Mathematical Society of Japan, which was held at Tokyo Imperial University. He proposed a new field theory of nuclear forces and also predicted the existence of the meson.

Sumi urged Yukawa to finish an English version of the paper quickly, which he did by the end of the month and submitted to the Society for publication.[14] His wife was an important part of his life, helping to manage his feelings, to empathize with his anxieties about the progress of his work, and to encourage his ambitions. He gained his very name from his wife's family, and Sumi was keen to have that name accorded the esteem she felt it deserved. The support of female family members was crucial to the public face that Yukawa and Tomonaga showed to the world.

Meson theory was developed in a number of papers in collaboration with Sakata, Taketani, and Minoru Kobayashi. In 1936, Yukawa was promoted to associate professor and in 1938 received his DSc. In 1938, Professor Kajurō Tamaki passed away. Yukawa returned to Kyoto Imperial University to replace him in 1939. By this time, his work was becoming known internationally and he was invited to attend the Solvay Conference on Elementary Particles and Their Interactions that year. Unfortunately, on his way there, the meeting was cancelled due to the imminent outbreak of World War II. Instead, he decided to go to the United States where he visited a number of physicists including Einstein. Yukawa had his first taste of policy-making as a member of the National Research Council (NRC) from 1939. In 1940, he received the Imperial Award of the Imperial Academy of Japan and became formally affiliated with Riken. He held an additional post of professor at Tokyo Imperial University during 1942–1946, giving lectures there from time to time. The worsening of the Allied bombing of Japan meant that traveling to Tokyo became more hazardous, and around 1943, he decided to confine himself to teaching at Kyoto. That year he received the Order of Cultural Merit.[15] Yukawa was also a member of the Chemical Research Institute, Kyoto Imperial University, from 1943 onward.[16]

At the end of World War II, in an effort to publish important Japanese research that had not yet been seen in English, Yukawa founded the journal *Progress of Theoretical Physics*, the first issue of which appeared in July 1946. Minoru Kobayashi assisted him in this project.[17] Yukawa was now regarded highly by the scientific community and this was reflected in his being made a member that year of the Imperial Academy, later known as the Japan Academy. Yukawa described Japan's postwar agenda in the following way at the inauguration ceremony of the Renewal Committee for Scientific Organization, of which he was a member:

> Making every effort to tide over the serious economic cricis [*sic*] on one hand, Japan is now on the way to accomplishing lots of hard work: such as we must establish a democratic nation, we must stabilize the national life, and further we must regenerate as a member of the international society.
>
> . . . it goes without saying that the freedom of research must be guaranteed in order to give full sway to the originality of researchers. But at the same time, scientific researches, whether the researcher is conscious or not, are always social phenomena.
>
> As long as each field of human knowledge . . . is isolated from each other, and its social life is disregarded, its true worth will not be discovered. The Renewal Committee should endeavor to establish an ideal organization. . . . Such effort is expected to be contributive not only to the promotion of the scientific researches in our country but also to the preparation for the day when Japan can participate in and cooperate with the international scientific researches as a member of the international society after the peace treaty is concluded.
>
> Finally I hope that people in general, especially the Government authorities concerned will fully understand that this work of the renewal of scientific research organization does not concern simply with the interests of the scientists, but it is an important problem of the nation most closely connected with the stabilization and improvement of the national life and will give us whole hearted support.[18]

Yukawa was soon able to rejoin the international community as one of the first Japanese scientists in a program of "Utilization of Japanese Scientists by the United States." It was envisaged by Harry Kelly

(Economic and Scientific Section, ESS) that other outstanding scientists would be encouraged to go to the United States rather than to "unfriendly" nations.[19] Yukawa arrived in San Francisco on September 3, 1948, en route to Princeton where he would take up Robert Oppenheimer's invitation of a visiting professorship at the Institute for Advanced Study, of which he was director (figure 5.1).[20] Speaking to journalists at San Francisco who were gathered to welcome the first Japanese scientist to visit the United States since before the war, Yukawa emphasized that his research would have nothing to do with the atomic bomb.[21]

Yukawa was feted in the United States. On April 28, 1949, it was announced that he would take up an appointment to the physics staff at Columbia University for the academic year 1949–1950, with assistance from the Rockefeller Foundation,[22] and shortly after on May 3, he was elected foreign associate to the National Academy of Sciences in Washington, D.C., the first such honor for a Japanese (figure 5.2). On November 4, Yukawa appeared on the front page of *The New York Times*. A smiling Yukawa was shown being congratulated by President Dwight D. Eisenhower, after it had been announced that he was the first Japanese to receive a Nobel Prize.[23] On his return to New York from Stockholm, where he received the Prize, he announced that part of the $30,000 prize would go toward a new institute of theoretical physics at Kyoto University.[24]

The awarding of the 1949 Nobel Prize in physics to Yukawa was the climax of many honors and served to bolster the confidence of the Japanese in themselves. In his absence, Yukawa was elected to the first term of the Science Council of Japan. The accolades continued with Yukawa being made emeritus professor of Osaka University in 1950, at the young age of 43. In July 1952, a new inter-university institute, Yukawa Hall, started operating to commemorate Yukawa's winning of the Nobel Prize. Upon his return from Columbia University in 1953, Yukawa was installed as director of the institute, which would be renamed the Research Institute for Fundamental Physics, housed in the building that would still be called Yukawa Hall.[25] Physicists jokingly referred to the building as "Yukawa Shrine," reflecting the revered status that Yukawa had achieved.[26]

II NAKASONE AND YUKAWA

In April 1953, Nakasone was reelected for the fourth time with Prime Minister Shigeru Yoshida still in government. An opportunity for further overseas travel arose on July 1 when he visited Harvard

Figure 5.1 Robert Oppenheimer, Hideki Yukawa, and Sin-itirō Tomonaga, November 1949, at the Institute of Advanced Studies, Princeton, just after the announcement of Yukawa's Nobel Prize.

Courtesy, AIP Emilio Segrè Visual Archives, Yukawa Collection.

Figure 5.2 Hideki Yukawa surrounded by reporters on the occasion of his return to Japan after receiving the Nobel Prize, 1950.

Source: Mainichi Shinbun-sha. Courtesy, AIP Emilio Segrè Visual Archives, Yukawa Collection.

University to attend a summer seminar program on global problems.[27] He took this opportunity to go on a fact-finding tour of the United States, to inspect facilities for the development of atomic energy. Impressed by how both the government and the private sector were pursuing nuclear power, he became determined to work toward the establishment of a national policy in Japan for the establishment of a system to promote atomic energy.[28]

In April 1953, the Arms Production Law had been promulgated and the Peace Preservation Agency announced its five-year-plan in September. Despite the "demilitarization" of Japan during the Occupation, 80–90 percent of Japan's armament production capacity was intact at the beginning of the Korean War. In 1952, a joint business and military committee for defense production was established to institutionalize the links between the military and industry. It was thus to be expected that in early 1953, Japanese business organized for rearmament. In October 1953, the Arms Production Council was started within Ministry of International Trade and Industry (MITI), under which eight subcommittees were created the following March to look into different technical aspects. On November 14, 1953, Kimura, the director of the Peace Preservation Agency, stated that he had hopes of expanding the Technical Research Institute and including in its research activities radio-guided missiles and atomic energy. It seems that when appropriate, the development of atomic energy for peaceful purposes and its development for bombs was interchangeable.[29] At the end of 1953, a Guided Missile Forum was jointly established by the Arms Production Committee, Arms Industries Association, and the Aeronautic Industries Association. In March 1954, the Mutual Security Agreement was signed, under which Japan was to receive weapons and other forms of aid. In May, the Defense Secrets Protection Law was established, followed by the creation in June of the Defense Council.[30] In 1954, however, arms production still accounted for a low 3.9 percent of machine industry production.

In 1954, Japanese business lobbied for research into guided missiles, and in the following year, orders were obtained for the production of jet fighters.[31] In July 1954, a Guided Missile Research Committee was established within the Peace Preservation Agency. From that September onward, the Agency became the Defense Agency.[32] It was thus only two years after the end of the Occupation that Japan had acquired a new army, navy, and air force. This was courtesy of legislation such as the Self Defense Forces Law of June 1954, and the law concerning the structure of the National Defense Council in July 1956.[33] It is against this backdrop that Nakasone embraced nuclear power.

The First United Nations International Conference on the Peaceful Uses of Atomic Energy, held in Geneva in August 1955, had a major impact on those who attended and recommendations were made by four Diet members to enact a basic atomic energy law and set up a Science and Technology Agency, as well as two public corporations to administer and conduct R&D for atomic energy. Among the four Diet members who attended the Geneva conference was Yasuhiro Nakasone. Nakasone was the delegation leader. This group of four, dubbed the "Four Samurai," visited the United States on their way back to Japan, and inspected research facilities such as the Argonne National Laboratory.[34] Almost every night of their trip, the four would discuss atomic energy legislation and ways in which an appropriate system could be put into place.[35] This cross-factional cooperation led to the formation of a Joint Diet Atomic Energy Committee, which was set up in October 1955 with these four politicians amongst its twelve members. Its chairman was Nakasone. The Committee was active in pushing for an atomic energy program. On November 14, 1955, a U.S.–Japan Atomic Energy Agreement was signed in Washington and, as a result, there was a great upsurge of interest amongst Japanese industrial circles in the commercialization of atomic energy.[36]

In November 1955, the third Ichirō Hatoyama government was formed. The business world was convinced that the conservative parties should join forces to form a strong, stable government in order to fight the increasing threat posed by the opposition and to implement solid economic policies. The Liberal Party (which had long supported Shigeru Yoshida's policies) and the Japan Democratic Party almagamated to form the Liberal Democratic Party (LDP), a party that has since been characterized by its emphasis on economic development, its close relationship with big business, and cooperation with the United States[37] This has helped to shape a policymaking structure that has been described as the "1955 system," a system lasting until the early 1970s, which saw close linkages between business and the LDP, electoral invincibility, clear policy directions in foreign policy and economic policy, and relative unity amongst the conservatives.[38] Nakasone was a rising star, and was elevated to the position of deputy secretary-general of the new Liberal Democratic Party.[39]

Nakasone was instrumental in having Matsutarō Shōriki appointed Minister of State in charge of atomic energy through his acquaintance with Bukichi Miki, a close adviser to the prime minister. On November 26, 1955, Nakasone and the Joint Diet Atomic Energy Committee made representations to the Cabinet Preparatory Council for the use of atomic energy. It was argued that there was an urgent need for a national policy for atomic energy. Nakasone was supported by three scientists who were present: Seiji Kaya, Yoshio Fujioka, and Sin-itirō Tomonaga. Formal government authority came to rest in the prime minister's office in the form of the Japan Atomic Energy Commission (JAEC), which was established under a special law on December 19, 1955. Nakasone made sure that the Commission was not an impotent one by having included a section that compelled the prime minister to respect the decisions of the Commission. It seems that such strong wording was unusual in legislation at the time.[40] The first meeting of the JAEC was held on January 4, 1956 and included physicist-members Hideki Yukawa and Yoshio Fujioka.[41]

Nakasone had made overtures to Kaya to encourage the Nobel-Prize winner Yukawa to become a member of the JAEC. Yukawa at first declined but agreed after reassurances to the effect that Yoshio Fujioka would take on the burden of any work that was generated. Kaya had previously hoped that Fujioka, chairman of the Science Council Committee for Atomic Energy Problems, would gain membership anyway. Kaya put it to Nakasone that Yukawa would only join on the condition that Fujioka too was included as a member.[42] Big business would be represented by Ichirō Ishikawa, president of Keidanren, and Hiromi Arisawa would also be a member.

In a 1956 article entitled "Japan Speeds Organization of Nuclear Program" in *Nucleonics*, MIT physicist and reactor specialist Clark Goodman reported on how atomic energy had become a major instrument for international cooperation and negotiation between Japan and the outside world. Within Japan, atomic energy was also used as a "powerful political propellant" in the academic world, industrial sector, and government circles. A major part of the activity involved the establishment of a broad base in nuclear science and engineering. The government tended to take the initiative in nuclear research whereas private industry largely took responsibility for development. Physicists were an important group in these activities.[43]

Yukawa, like Werner Heisenberg in Germany, was a winner of the Nobel Prize for physics and director of a research institute. As such, they both represented a combination of scientific prestige and academic power.[44] In Yukawa's case, this was enshrined within the Research Institute for Fundamental Physics and its journals—the gatehouse to physics in Japan. The Elementary Particle Theory Group provided Yukawa with organizational support, a veritable national network of physicists.

The JAEC gave Yukawa another forum, but this was short-lived. With the atomic energy law of 1956, various organizations were created to "bring" atomic energy to Japan in a peaceful form. The JAEC reported to the prime minister through the chairman Shōriki, who also was a minister in the cabinet. Other members (two full-time and two part-time) were appointed by the prime minister. The full-time members were Ishikawa [also chairman of the board of directors of the Japan Atomic Energy Research Institute (JAERI)] and Fujioka (physicist and formerly Tokyo University professor). Yukawa and Arisawa (Tokyo University economics professor) were part-time members. By his appointment, Yukawa was made an unwilling accomplice to state strategies for atomic energy. Yukawa's decision to join the Commission was prompted partly by the example of the Indian particle physicist Homi J. Bhabha, who combined basic research with application in his three roles as director of a research institute for fundamental physics, director of the atomic energy agency, and chairman of his country's Atomic Energy Commission. Yukawa hoped that he could provide such a bridge between academia, government, and business.[45]

An Atomic Energy Bureau was created in Japan to provide administrative support for the JAEC, a JAERI to carry out related R&D, and the Joint Diet Committee on Atomic Energy to help politicians monitor developments in nuclear power. The chairman of the latter was Nakasone. A Nuclear Fuel Corporation was formed to take responsibility for prospecting, mining, refining, and administering nuclear fuel. The Japan Atomic Industrial Forum was established on March 1, 1956 to represent around 300 companies interested in atomic energy.[46]

Yukawa hoped to exert considerable influence as a member of the Commission, but Chairman Shōriki had firm ideas about the direction of the JAEC. His was a formidable character. Like many of Japan's elite bureaucrats, Shōriki was a law graduate of Tokyo Imperial University. He spent 13 years on the Metropolitan Police Board before serving as chief of the Criminal Affairs Bureau until 1924.[47] He later became president and owner of the *Yomiuri shinbun* newspaper company, boosting its circulation and making it the third best-selling newspaper in Japan. During the war, the *Yomiuri* was a fervently nationalist newspaper and reflected Shōriki's support of the military order.[48] He was held in Sugamo prison as a suspected war criminal for 21 months, from December 12, 1945 until September 1, 1947, during which time he was interrogated twice. In August 1951, he was officially de-purged. In late 1955, Shōriki was made a cabinet minister by his friend, Prime Minister Ichirō Hatoyama. On January 1, 1956, he became Japan's first Atomic Energy Commissioner.[49] What outraged physicists such as Sakata was that Shōriki ignored any lessons from

Japan's defeat and continued about his business as usual. His enthusiasm for atomic energy was not tempered by any wartime guilt.

The JAEC officially began on January 1, 1956, but members were soon startled by a statement on January 5 by the Chairman Matsutarō Shōriki, the day after the first meeting of the Commission. Shōriki issued a surprise announcement that Japan would sign an atomic energy agreement with the United States, and attempt to achieve power generation within five years. Commission members were concerned that Shōriki would use the JAEC to merely rubber-stamp his ambitious plans. Yukawa was more guarded about the question of the importation of power reactors, but Shōriki wanted to speed up the entire process of atomic energy development, including the generation of power. Yukawa and Hiromi Arisawa were forced into becoming watchdogs over government and business. They both argued that the development of atomic energy should proceed only after basic research had been conducted, the appropriate personnel trained, and the necessary technology developed.[50] Yukawa threatened to resign. The Commission attempted to defuse the situation by releasing a more modest statement, which watered down Shōriki's words.[51]

Yukawa had, from the time of Shōriki's controversial statement, felt that taking on membership of the JAEC had been a mistake. He was persuaded by Nakasone and others not to resign. He was told that if he were to do so, it would cause a great calamity, attract adverse media attention, and that even the cabinet could be put into danger. Yukawa gave into such representations for the time being.[52]

For Yukawa, some of the JAEC discussions must have been quite tedious even though he attended only one in two meetings and was often away because of overseas travel and ill health.[53] But the biggest shock for Yukawa was the politicized nature of all that the Commission did. Everything from the budget, government bills, and JAERI site selection involved political considerations. From April 1956, the basic planning for the development of atomic energy began in earnest. Yet, in late April, Yukawa indicated his desire to resign from the JAEC. The letter to Shōriki cited poor health and interference with research as reasons for his resignation, but there was concern that the running of the Commission and problems associated with the proposed introduction of reactors from abroad were the real reasons. The minister responded to the press with the throwaway-lines that

Yukawa's intention to resign was, in a way, apparent from the time he became a member of the Atomic Energy Commission, so it's nothing to make a fuss about.[54]

Despite such public utterances, the Nobel Prize–winning Yukawa was considered a "gilt signboard" for the JAEC. Coming only four months after the formation of the Commission, his resignation was considered a major blow to the government. Yasuhiro Nakasone apparently acted on Shōriki's behalf and negotiated with Yukawa to have him stay on for one more year.[55] Yukawa continued for the time being but there was a general feeling that he would not last his full term as member of the Commission.[56]

Foster Hailey, a journalist writing for *The New York Times*, saw the tensions not only in terms of the professors wishing to use Japanese talent and resources to develop their own infrastructure for nuclear power, but also as a reflection of "whether atomic energy should be developed primarily by the Government or by 'speculative enterprises.' Mr. Shōriki has made it plain he favors the latter course." Hailey went on to report that "Mr. Shōriki appears to favor the most rapid possible development by Japan of economically feasible atomic power facilities." As for who would prevail in the controversy, Bailey had little doubt that "In any disagreement between Mr. Shōriki and the four professors, the odds would favor Mr. Shōriki. He is a Minister Without Portfolio in the cabinet of Ichiro Hatoyama and apparently has the full confidence of the Premier."[57]

Although Yukawa did not have a major impact on government plans for nuclear power, the press considered his presence on the JAEC as an important voice of caution. His departure was seen as strengthening the position of atomic energy advocates in government and business who were pushing for rapid development using foreign technology.[58] There was also the worry that Japan's atomic energy program might lose esteem abroad if its association with Yukawa was lost. As Yukawa's longtime friend Yūichi Yuasa commented at the time, "He is not only Japan's Yukawa, he belongs to the world."[59]

On March 18, 1957, almost one year after his conversation with Nakasone, Yukawa finally resigned from the Commission, citing ill health. It was sufficiently newsworthy to even be reported in *The New York Times* the day after.[60] He had spent a little over a year on the Commission. The press argued that although the government had assembled a fine array of members, their talents were under-utilized. The government considered the Commission an advisory body, and neglected to pay serious attention to its decisions. The press reported that if the Commission was not careful, it would merely become a means of legitimating government policy.[61]

Yukawa's resignation brought to an end his special role as the government's nuclear adviser. To some extent, he was replaced by physicists such as Ryōkichi Sagane and Seishi Kikuchi,[62] who were more amenable to government policy. Yukawa pursued activities to which he felt more committed, traveling overseas to attend the first Pugwash conference and receiving an honorary doctorate from the University of Paris. His resignation from the JAEC did not preclude his attendance at the Second International Conference for the Peaceful Use of Atomic Energy in 1958. Overseas trips continued with a visit to Kiev in 1959 to attend an international conference on high-energy physics.[63]

Although considered Yukawa's equal in many ways, it wasn't until 1965 that Tomonaga joined Yukawa in the exclusive club of Japanese Nobel laureates. But long before, Tomonaga had been considered Nishina's successor, blessed with an affability and talent to nurture students, which Yukawa lacked. This ability to work with others made up for Yukawa's aloofness. In the Science Council, which he became president of, it would be put to good use.

III SIN-ITIRŌ TOMONAGA

Sin-itirō Tomonaga came from a highly cultured family, his gift for writing almost as renowned as his physics.[64] His family background is further evidence of how elite physicists came from privileged classes. He was born less than a year before Yukawa, on March 31, 1906, the eldest son of Sanjurō Tomonaga and his wife Hide. Hide was the daughter of a retainer of Suo-no-kami Matsudaira, the leader of a feudal clan. Her mother was a member of the Okano clan.[65] Hide was a graduate of Ochanomizu Girls Middle School. Sanjurō was of Nagasaki samurai background, born in 1871, the third son of a former Ōmura clansman (daimyō retainer) named Jinjirō Tomonaga. Sanjurō studied philosophy and history of philosophy at Tokyo Imperial University and at the time of Sin-itirō's birth, was professor of philosophy at Shinshū University (later known as Ōtani University) in Tokyo. The following year, Sanjurō took up the position of associate professor at Kyoto Imperial University and in 1909, decided to go abroad for further study, leaving his wife and child in Tokyo.[66] Separated from his father for many years, Sin-itirō grew up for some time only knowing his mother. Yoshio Nishina would later provide a father figure for the mentor-less young man.

Sanjurō visited Germany, France, and England, spending most of his time at Heidelberg University. Upon his return to Japan in 1913, the family took up residence again in Kyoto, where he was made professor (retiring in 1931 and passing away in 1951). One of Sanjurō's eminent colleagues at Kyoto was the philosopher Kitarō Nishida. In 1918, Sin-itirō entered Kyoto Prefectural First Middle School, a prestigious school that often led to the Third High School, Kyoto Imperial University, and sometimes public intellectual life. Due to ill health, he started school one term late. Nishina, too, had missed long periods of study due to illness, as did the sickly Taketani. Their lives combined both periods of power and powerlessness. Tomonaga would overcome his weak body and prove himself through academic achievement, emulating his previously absent father. Fortunately for physicists, intellect and rationality were equated with manliness, not physical fitness.

In 1923, Sin-itirō proceeded to the Third High School. That year, Einstein visited Japan, prompting Tomonaga to read books on physics by people such as Jun Ishiwara. While at high school, Sin-itirō had the opportunity to visit Hantarō Nagaoka and speak with him regarding physics, thanks to the friendship between Nagaoka's wife and Sin-itirō's mother.[67] In this small way, female networks provided advantages for the children of academics.

In 1926, Tomonaga gained admission to the Faculty of Science at Kyoto Imperial University, majoring in physics. He and Hideki Yukawa chose to specialize in quantum mechanics in their third year in 1928. Although both Tomonaga and Yukawa were formally attached to the lab of Professor Kajurō Tamaki, the lack of an appropriate supervisor meant that the students studied from books by themselves. Tomonaga graduated in 1929 and chose to remain at university as an unpaid associate with Yukawa. As both were from well-to-do, highly educated families, this was not seen as a problem. It is not surprising that class hierarchies served to propel the two to positions of considerable influence, free of the burden of having to provide an income for an extended family.

Tomonaga traveled to Tokyo in September 1929 to attend lectures given by Heisenberg and Dirac at Riken. These lectures, and an inspiring special lecture on quantum mechanics given at Kyoto by Nishina in 1931, prompted Tomonaga to join the Nishina lab at Riken in 1932. After an initial provisional period of three months, Tomonaga was formally made a Riken researcher. Sakata followed in his footsteps in 1933. By this time, a small group of young theoretical physicists had gathered at Nishina's lab. Nishina provided them with a father figure. He was a mentor for the younger generation and helped to bring them to intellectual and social power.

With Sakata's move to Osaka University in 1934, his Kyoto University classmate Minoru Kobayashi came to replace him at the

Nishina laboratory. Hidehiko Tamaki (Tokyo Imperial University) also joined the lab. Tomonaga spent the 1935 summer at a mountain retreat in North Karuizawa, co-translating Paul Dirac's *Quantum Mechanics* with Kobayashi and Tamaki, after a request from Nishina.[68] Tomonaga studied nuclear theory at Leipzig University under Werner Heisenberg from June 1937 to August 1939, thanks to a researcher-exchange agreement between Riken and Leipzig University. One can understand the strong bonds between Japanese teachers and students by looking at the diary that Tomonaga kept while in Leipzig. The date is November 22, 1938. Tomonaga had just received a letter from Nishina, which brought tears to his eyes. Knowing that Tomonaga was depressed about his lack of progress in his work, Nishina tried to cheer him up by writing that

> It is a matter of luck as to whether or not one will achieve anything. We stand at the crossroads, not knowing what lies before us. Whether we go left or right depends upon luck and our feelings at the time. Even if there is a big difference if one proceeds along a certain path, there's not much point worrying about that. In time, one will get lucky and things will go right. I always keep that in mind and hope for the unexpected. Anyway, cheer up and take care. All one can do is to try to make sure that luck comes one's way.[69]

Tomonaga would not forget Nishina's words and was grateful for the concern shown by him. It was a relationship not unlike that of father and son. Unlike Sakata and Taketani who felt that Marx and Engels offered some methodology and blueprint for their life and work, to Nishina and his student Tomonaga, it seemed that it was largely a matter of luck and that one just muddled along only hoping for the best.

Another Japanese physicist studying at Leipzig was Satoshi Watanabe. Japanese predecessors at Leipzig included Seizō Kiuchi, Takuzō Sakai, Seishi Kikuchi, Yoshio Fujioka, Kanetaka Ariyama, and Kiichirō Ochiai. While Watanabe and Tomonaga were studying there, Kobayashi took up a position at Osaka University. Tomonaga returned to Japan in 1939, meeting Yukawa on the same ship. Watanabe returned slightly later.[70] Yukawa had planned to attend the Solvay Conference but the meeting was cancelled. Upon his return, Tomonaga received his DSc from Tokyo Imperial University and married Ryōko Sekiguchi the following year. His new wife was the daughter of R. Sekiguchi who had been director of the Tokyo Metropolitan Observatory.[71] From June 1940, Tomonaga was a Riken research fellow, as well as being a professor at the Tokyo University of Science and Literature from 1941.

From around mid-1943, Tomonaga was mobilized to carry out research on magnetrons and ultra-shortwave circuits at the Naval Research Institute laboratory at Shimada, along with Masao Kotani, Yūsuke Hagihara, Yuzuru Watase, Tatsuoki Miyazima, and others. It appears that Nishina was responsible for Tomonaga going to Shimada. Tomonaga had attended magnetron study meetings in the Department of Physics at Tokyo Imperial University and developed an interest in magnetrons. It was suggested that he go to Shimada to continue that research. Osaka's Yuzuru Watase asked Minoru Oda (then at Seishi Kikuchi's lab at Osaka University) to join him in helping with experiments on a new type of magnetron. Tomonaga, Kotani, and Hagihara were already there. Oda, Watase, and others went about making various types of magnetrons at Shimada.[72] In 1944, Tomonaga was appointed to a lectureship at Tokyo Imperial University. From August 1944 to July 1945, Tomonaga's work also included a temporary assignment to the Tama Army Technical Institute.[73]

In 1946, physicists and students from the Department of Theoretical Physics at Tokyo University of Science and Literature (also known as Tokyo Bunrika University) joined some of the physics students that Tomonaga had taught at Tokyo Imperial University to hold a seminar once a week in what was left of the University of Science and Literature campus at Ōtsuka. These seminars were known as the "Tomonaga Seminar" or "Friday Seminar." In 1946, Tomonaga won the Asahi Cultural Prize for his work on meson theory and his super-many-time theory. As word spread of Tomonaga's seminar, the number of students attending them rapidly increased. With Nishina devoting his energies to administration after the war, the leadership of the Tokyo group of physicists gradually shifted to Tomonaga. In June 1947, Tokyo University of Science and Literature's Department of Theoretical Physics and part of the Experimental Physics Department moved to the burnt-out remains of the Army Technical Research Institute at Ōkubo. The Tomonaga Seminar continued on strongly, always attracting around 30 young researchers from all parts of Japan, becoming a mecca for elementary particle theorists, an opportunity for male-bonding, and a rallying point for what would become known as the Elementary Particle Theory Group (figure 5.3).[74] It provided a space where bright, young men could exercise their social and intellectual power.

In 1948, Tomonaga and Masao Kotani were awarded the Japan Academy Prize for their joint research on magnetrons. Tomonaga was elected to the Science Council of Japan that year and, in 1949, was reappointed professor to what had become the Tokyo University of

Figure 5.3 Sin-itirō Tomonaga, center, with hand raised, seated beside Nobuyuki Fukuda at a seminar at the Okubo branch of the Physics Department, Tokyo Bunrika University, Tokyo, 1948.

Photograph: Shunkichi Kikuchi. *Source*: University of Tsukuba, Tomonaga Memorial Room. Courtesy, AIP Emilio Segrè Visual Archives.

Education. Upon Tomonaga's departure to the United States on August 9, 1949, on an invitation from Robert Oppenheimer to visit the Institute for Advanced Study, Princeton, the Tomonaga Group dispersed as well. Unlike the fanfare that accompanied Yukawa's departure, Tomonaga's passage was a much quieter affair, in a third-class cabin of a freighter.[75]

The Tomonaga Seminar came to an end but his students were not forgotten. Both Tomonaga and Yukawa had been impressed by the intellectual environment of the Institute for Advanced Study, which supported a combination of faculty and visiting scholars. Although the Institute was privately endowed, Tomonaga and Yukawa entertained hopes of recreating that environment back home in Japan. The awarding of the 1949 Nobel Prize for physics to Yukawa provided the catalyst for this in the form of government support for the establishment of Yukawa Hall at Kyoto University.[76] Tomonaga's Nobel Prize would be much later in coming but he was, nevertheless, held in high regard by the scientific community.

Tomonaga's year at Princeton came to an end in the summer of 1950. He returned to Japan a new man, wearing fashionable suits made in the United States, and smiling with a new set of false teeth.[77] His new appearance gave Tomonaga added confidence, and it was perhaps just as well. With the death of Nishina in 1951, much responsibility for nuclear physics fell on Tomonaga's shoulders. It has been suggested that Tomonaga became involved in science policy out of a sense of duty and gratitude to Nishina.[78] He took over from Nishina as chairman of the Japan Science Council (JSC) Liaison Committee for Nuclear Research [later known as the Special Committee for Nuclear Research (SCNR)] and busied himself with an agenda of lobbying for the establishment of the Cosmic Ray Observatory at Mt. Norikura as a joint-use facility (but formally attached to Tokyo University), and reviving cyclotron research. Ernest Lawrence visited Japan in 1951 and with his encouragement, construction of a new, small, 3.7-MeV cyclotron began at the Scientific Research Institute (formerly known as Riken) in 1951. It was completed in 1952. Funds for the cyclotron were provided by the Ministry of International Trade and Industry. Not to be outdone, the Ministry of Education helped finance a cyclotron at Osaka University (13 MeV) and the private sector supported construction of a machine at Kyoto University (14 MeV).[79] With Tomonaga's coming of age as a scientific community leader, he was made a member of the Japan Academy in 1951 and awarded the Order of Cultural Merit in 1952.[80]

Tomonaga, had feelings similar to those of Yukawa regarding the rushed development of atomic energy. His famous words were that

> The promotion of research into atomic energy is like trying to build the second storey before the foundations are firm.[81]

In his view, nuclear physics research should provide that base. The Research Institute for Fundamental Physics and the Mt. Norikura Cosmic Ray Observatory were viewed as low-cost ways by which the common-use research institute concept could be started.

In 1953, Tomonaga was appointed adjunct professor at Kyoto University. That year, he was involved in discussions with Seishi Kikuchi, Yoshio Fujioka, and others regarding the establishment of an Institute for Nuclear Study (INS).[82] Physicists were particularly interested in promoting basic research. In nuclear physics, the increasing scale of research demanded changes. This was a combination of rapid advances that had been made in the field as well as the accompanying desire for larger apparatus. It was felt that the existing system of research in Japanese universities was unsuited to the new needs of experimental nuclear research. The creation of the INS was important in helping to establish the concept of national joint-use research institutes in Japan. The INS, with its inaugural director, Seishi Kikuchi, was one important early example of physicists' desire to shape a new kind of research system.

At various meetings of the Physical Society of Japan, discussions were held and the SCNR under Tomonaga's chairmanship, formulated plans. It was felt that physicists could help encourage democracy in Japan by adopting an infrastructure with democratic principles underlying it. Nuclear physicists organized themselves into a lobby group called the "Genshikaku Danwakai" ("Association of Experimental Nuclear Physicists"), inspired by the Elementary Particle Theory Group, to facilitate the establishment and development of the proposed Institute. In May 1953, the Science Council of Japan recommended the establishment of INS to the Japanese government. The Science Council hoped to open up the research facilities to all researchers throughout the country and to conduct research programs that would be administered by an inter-university "democratic" committee. It was also hoped that there would be frequent exchange of physicists between INS and other universities. Tokyo University would not agree with the idea of an inter-university steering committee, lest the autonomy of the university council and faculty be threatened. They did agree with the common-use concept and after considerable negotiation, the physicists became resigned to the fact that although informally inter-university, formally it would have to operate as a research institute of Tokyo University.[83]

The Genshikaku Danwakai put together initial plans for personnel and experimental apparatus in January 1954 and an ad hoc committee was formed to help preparations, first meeting in May 1954. Unfortunately for the nuclear physicists, 1954 was a controversial year for anything connected with atomic energy. In March, the Japanese government boldly went ahead with its first budget allocation for atomic energy without proper consultation with physicists or the general public. At the same time, the United States began H-bomb tests in the Pacific. The SCNR, via the Science Council, issued a declaration that all atomic energy be in accordance with the three basic principles of non-secrecy (free access), democracy, and independence. In July 1954, a site in Tanashi, Tokyo, was selected for the Institute, but local inhabitants took fright and protested against its construction, fearing that academic research might be transformed into research for military use. Tomonaga and other physicists (belonging to SCNR, INS representatives, and Genshikaku Danwakai) tried to alleviate local fears and persuade residents that there was no radiation hazard. Yet, given the enthusiasm with which the government embraced atomic energy, they could not guarantee the outcome. After lengthy discussions with the townsfolk for over 6 months, INS was formally established in July 1955 at Tanashi, some 25 miles from central Tokyo.[84]

The agenda of INS was to promote research in nuclear and particle physics. This included low-energy physics, high-energy physics, cosmic-ray physics, and theoretical physics. Although it was formally attached to Tokyo University, the facilities were open to researchers throughout Japan. It was thought that a high-energy accelerator would be too costly and technically difficult; hence, they decided to opt for a low-energy accelerator. The decision was made to establish an INS based around a cyclotron. Thus, one of the first projects was the construction of a 160-cm, 55-MeV, variable energy frequency-modulated (FM) cyclotron, which was completed in 1957. As cyclotrons had already been constructed at Osaka and Kyoto Universities, it was decided that the INS cyclotron should be both fixed frequency (FF) and frequency modulated (FM cyclotron or synchro-cyclotron). A 750-MeV (later boosted to 1.3 GeV) electron synchrotron was commenced in 1956 and completed in 1961 as the first accelerator for elementary particle physics in Japan and as preparation for a high-energy accelerator.[85] In the late 1960s, INS was the home of various preparatory studies for the 12-GeV proton synchrotron, which would form the core of the KEK High Energy Physics Laboratory at Tsukuba.[86]

Following Sakata's example in the Department of Physics at Nagoya University, the INS physicists attempted to abolish the "feudalism" of the laboratory and set up a democratic base for the administration of research activities in which all researchers throughout Japan had a voice. Unfortunately, this "democratic structure" relied on the goodwill of the INS faculty board, which could ignore it if it so wished. Like the example of the Science Council of Japan, which represented scholars from throughout Japan, the government was under no obligation to follow their advice.

Tokyo University refused to hand control of INS administration to an inter-university committee of researchers, but rather left decision-making with the INS faculty board. The SCNR overcame this impasse by setting up an inter-university General Committee with membership being equally divided between INS members and non-Tokyo University representatives. Although formally there was no compulsion on the faculty board to heed the recommendations of the SCNR General Committee, it was felt that there was a "moral obligation" on their part to do so. Another aspect of the democratization of INS was that the faculty board (consisting of professors and associate professors) would also receive input from an INS staff steering committee, which in turn received input from a larger researcher's "congress" made up of INS staff. Furthermore, each section (low-energy physics, high-energy physics, cosmic-ray physics, and theoretical physics) held weekly staff meetings to discuss issues of concern. To encourage the idea of common-use of facilities by other researchers, special committees of researchers from each field (chosen from throughout Japan) were formed with representatives also coming from the inter-university General Committee.[87]

The democratization of research facilities for physicists identified other problems such as the wide discrepancy in status between researchers compared to technicians. Also, as INS formally belonged to Tokyo University, only that university's students could be educated there. Another problem associated with the democratic use of one accelerator by all nuclear physicists in Japan was that machine time for experiments was short. Physicists attempted to overcome this problem by assigning longer periods for projects and to lobby for a second INS.[88] But democratization of the research system did not come easy, and over the years, these issues would be raised again with the establishment of research institutes in other universities.[89]

Tomonaga's involvement in developing research infrastructure also included membership of the management committees of the Research Institute for Fundamental Physics and the Cosmic Ray Observatory. In 1955, the cyclotrons that Tomonaga had worked to revive were rebuilt. Tomonaga's activities became increasingly administrative, especially with his appointment in 1956 as president of Tokyo University of Education, a job that he would continue until 1962. In 1959, he was selected for membership for the new Science and Technology Council (the government's rival to the Science Council), while still maintaining close ties with the Science Council through groups such as the SCNR Future Plans for Nuclear Research Subcommittee. In 1960, he was re-elected University President, made a Riken consultant, and Japan UNESCO Committee member. In October 1961, he attended the Solvay Conference in Brussels. In 1963, he was made the director of Tokyo University of Education's Institute for Optical Research, research adviser ("gakujutsu komon") to the Ministry of Education, and president of the Science Council of Japan. Tomonaga would remain as JSC president until 1969.

Tomonaga's presidency of the Science Council coincided with a period of conflict between academics and government–industrial circles, and scientists and technologists. This conflict was manifested in the activities of the Science and Technology Agency and the creation of the Science and Technology Council.

IV   NAKASONE AND TOMONAGA

Nakasone had become convinced of the need for a science and technology agency. One of the other "four samurai," Diet member Masao Maeda, spearheaded preparations for the establishment of an agency, but Nakasone was able to have considerable input. According to his view of things, the administration of atomic energy should remain with the agency by having the secretariat (Atomic Energy Bureau) for the Atomic Energy Commission located within the Science and Technology Agency.[90] So it was that in May 1956, the Science and Technology Agency was established (see figure 5.4), headed by none other than Shōriki, who was also a state minister and chairman of the Atomic Energy Commission. Shōriki had also been

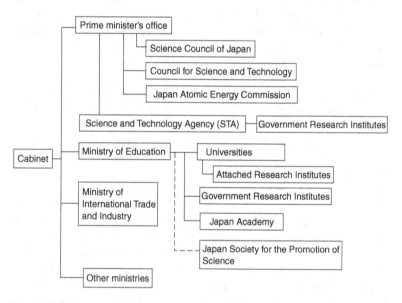

Figure 5.4 Major elements of government R&D policy-making structure, ca. 1950s–1960s.

Adapted from Ministry of Education, Science and Culture, *An Outline of the University-Based Research System in Japan* (Tokyo, 1983), p. 3.

one of the principal persons behind the creation of the Japan Atomic Industrial Forum in March 1956.[91]

One year later in May 1957, Nakasone accompanied Nobusuke Kishi, who had just become prime minister, on a tour of Southeast Asia, the Middle East, and Eastern Europe, as part of his entourage. Come July, Nakasone became a member of the Cabinet Committee for examination of the Constitution, one of his pet interests. In May 1958, he was reelected for the sixth time as part of the Kishi government and in June, became a director of the Liberal Democratic Party and deputy chairman of the LDP Executive Council.[92] Nakasone's career was moving ahead very fast. He was making his interest in science and technology work to his advantage.

In June 1959, Nakasone became a state minister, director of the Science and Technology Agency for one year, and until July 1960, chairman of the JAEC. In November 1960, he was reelected (for the seventh time) with the Hayato Ikeda government. In January 1961, the Cabinet Committee for the Examination of the Constitution selected Nakasone for a fact-finding tour of North, Central, and South America. In August, he became chairman of the LDP Special Committee for Science and Technology. Nakasone thus emerged as a powerful politician, with a strong focus on science and technology, and a distinct liking for international travel. In November 1962, this even took him as far as Antarctica. Physicists feared that such interest might see Japan joining other nations in the race to carve up the last frontier and gain further territory.[93]

In January 1963, Nakasone established the League for the Promotion of the Public Election of the Prime Minister, and was chief of the promotional headquarters for the League's national movement. In November, he was reelected for the eighth time with the Ikeda government and in January 1964, became chairman of the LDP Committee for the Drafting of the 1964 Campaign Direction. Nakasone's interests extended to the arts, with chairmanship of the Diet member's League for the Promotion of the Arts from March 1964. This was in keeping with his self-image for he viewed himself as a writer. In March 1965, Nakasone became chairman of the sub-committee for Asia and Africa within the LDP Foreign Policy Study Group and the next month was appointed an adviser to Ambassador Kawashima who attended the tenth anniversary of the First Asia and Africa Council, which was held in Indonesia.[94]

In September 1965, Nakasone was made adviser to the Japanese representative to the Ninth General Assembly of the International Atomic Energy Agency. One of Nakasone's mentors, Ichirō Kōno, passed away in 1965. Several young politicians, such as Nakasone and Kiyoshi Mori, sought to take over from Kōno. Seishi Shigemasa, a close colleague of Kōno's, was chosen as new faction leader, but he was unable to control the faction. A board of five directors was instead elected to carry out the important decision-making for the faction, but this arrangement too proved unsatisfactory. The faction split in December 1966, with Nakasone inheriting many of Kōno's supporters and the others going to Mori. Now with his own faction, Nakasone sought to position himself to achieve his ultimate goal of the prime ministership of Japan. He attracted more financial backing and was able to slowly increase his faction membership.[95] From around this time the LDP increased its control over policymaking and its influence was said to rival that of the public service bureaucrats who had dominated the decision-making process for the first two decades of the postwar period.[96]

The JSC, meanwhile, became increasingly critical of government policy in science and technology, and increasingly threatened by organizations such as the government's own Science and Technology Council (see figure 5.4). Social issues during the 1960s were also a source of dissension, with the Cuban missile crisis, American intervention in Vietnam, American bases on Okinawa, and visits by American nuclear submarines to Japanese ports all providing points of conflict with the government.[97] The Science Council became wracked by internal conflict between left and right wing politics, the Establishment versus anti-Establishment, and basic versus applied research.

One major controversy surrounded the question of the use of Self-Defense Force frigates as vessels for Antarctic study missions. By signing the 1951 Peace Treaty, Japan renounced all claim to certain parts of Antarctica, which had been the result of a past expedition. Whether this included future claims was a matter for debate. Scientific research stations were established by Japan in 1957, 1970, and 1985. Any military involvement in the Japanese presence in Antarctica was deemed inappropriate for fear of accusations of territorial expansionism.[98]

Other problems to embroil Tomonaga were the relationship between the Science Council and the U.S.–Japan Committee on Scientific Cooperation, which had been established in 1961; and the chairmanship of a new committee called the "Soken Junbi Chōsa Iinkai" (SJC or here referred to as the KEK Preparatory Committee), which took responsibility for planning for the Research Institute for Elementary Particle Physics (later known as the KEK High Energy Physics Laboratory).[99]

In 1964, the deterioration of relations between the Ministry of Education and the Science Council became a subject of great concern for the SCNR and Tomonaga was in the thick of it. The overlap of membership in the Science Council and in Ministry of Education committees was seen as threatening the autonomy of the Council and leading to the decline of its influence in deciding science policy matters.[100] The Council for Research Institutes ("Kenkyūjo Kyōgikai") and the newly created Council for the Promotion of Science ("Gakujutsu Shōrei Shingikai") were cited as examples of where this was and would be happening. Tomonaga, who was thought of as one of the prime "culprits," performed the feat of juggling the presidencies of both the Science Council and the Council for Research Institutes. It was hoped that the view and interests of the Science Council would always be taken into account in any decisions that were made by the latter body. The complaint was made that the various committees of the Science Council were not kept informed of the activities of the Kenkyūjo Kyōgikai and that members would often have to seek information through private channels. Contact between the two organizations relied, to a great extent, on the coincidence that the president was the same for both.[101]

There was a very real threat that the Ministry of Education would slowly divest the Science Council of its powers and install them, one by one, in committees under its own jurisdiction. The Science Council sub-committee for budget policy was, as of April 1964, dissolved and in a way accommodated into the Council for the Promotion of Science, which was considered as potentially being a more powerful body.[102] That same year, Tomonaga became chairman of the board of directors of the Nishina Foundation and professor of Tokyo University's INS. These many appointments were capped by the Nobel Prize for physics in 1965, for his super-many-time theory, renormalization theory in quantum electrodynamics, and quantum-mechanical many-body problems.[103] Tomonaga was joint recipient with Julian Schwinger and Richard P. Feynman. The award of the Nobel Prize to Tomonaga affected the Japanese government's disposition toward plans for an Elementary Particle Research Institute, in the same way that Yukawa's Nobel Prize assisted plans for the Institute of Nuclear Study. The Nobel Prize is credited with assisting the Institute's committee in receiving funds for preparations for the proposed high-energy physics laboratory. The prestige of the Nobel Prize also served to deflect criticism of Tomonaga as JSC President.[104]

In 1967, a major issue that Tomonaga presided over was the question of U.S. Army funding of research activities in Japanese universities.

It had come to light that such funding had been received by the organizers of an international conference on semiconductors, and the Japanese were not at all used to any military–academic cooperation in research. There was considerable anti-American feeling in left wing intellectual circles at the time, and it would be decades before university–industry cooperation would be welcome let alone that with the military.

Tomonaga retired from Tokyo University of Education in 1969, as well as completing his term of office as JSC president. His retirement from public duties coincided with increasing antiscience feeling amongst the general public. He died in 1979 from cancer.[105]

V   Yukawa and Tomonaga Compared

In surveying the activities of Tomonaga and Yukawa, certain common characteristics emerge. Both physicists were the offspring of academics, and both followed an elite academic route via the First Middle School and Third High School to Kyoto Imperial University. Both also grew up during the liberal period after World War I known as Taishō democracy, retaining Eastern traditions and taking on Western influences as it suited them. In contrast, Shōichi Sakata, who was some four years their junior, grew up during a time when Japan was increasingly taking on a militaristic footing, when modern rationalism was in vogue and when access to quantum mechanics was much easier.[106] The Japanese scientific community was not insulated from the traditional society, nor the changes that impacted upon it. Various philosophies influenced physicists. They grew up in belief systems that were prevalent in Japan before World War II, sometimes actively taking up ideas that attracted them. For Yukawa, Marxism was of little interest unlike for Sakata and Taketani, for whom it was a strong influence in research and political activities. Yukawa attributes part of his success to a type of Eastern intuition.[107]

Yukawa contends that his conception of nature has been influenced by Taoism. For him, the Chinese classical knowledge of his childhood provided a source for the development of his scientific thought. He considered the Chinese and Japanese as having a type of intuition that is part of the "Oriental approach." This intuition was, for Japanese intellectuals, counterbalanced by rational philosophies such as Taoism. The Taoist sages Laotse and Chuangtse placed Nature, not humankind, at the center of things. Natural phenomena were viewed in terms of cause and effect.[108] Like a Chinese sage, Yukawa himself described the feeling of achievement at having arrived at a new theory in the following elegant manner:

> I felt like a traveler who rests himself at a small tea shop at the top of a mountain slope. At that time I was not thinking about whether there were any more mountains ahead.[109]

Although full of doubt as to where his path might lead, it is the landscape that controls him, not vice versa. In old age, Yukawa would grow a long, sage-like beard, in keeping with the persona that he had styled for himself from early on.

Yukawa argues that his theoretical prediction of the existence of the meson was facilitated by both intuitive aspects of Japanese thought and rationalistic analogies of Taoist philosophy. He feels that Chuangtse's notion of invisible moulds may have helped him to intuitively grasp the idea of the appearance and disappearance of different elementary particles.[110] This approach to nature reflects how Yukawa drew on the ideas of the quickly changing society into which he was born and socialized, but they also reflect how he consciously attempted to fashion himself as an elite scholar with strong ties to Japan's cultural past. By emphasizing how he appropriated and reworked cultural resources in his physics, he defined what it was to be a Japanese physicist. Philosophy provided Yukawa with his meta-narrative, just like Marxism was the answer for Sakata and Taketani (figure 5.5).

Tomonaga's cultural influences were more recent than Yukawa's and also more international in flavor. He refers to the influence of his father Sanjurō Tomonaga who was a philosopher at Kyoto Imperial University and one of the earliest Kantians in Japan. Tomonaga was also influenced by his father's colleague Kitarō Nishida, as was Yukawa. Nishida tried to develop a system of thought of his own in which there were two ways of knowing things: direct perception and knowledge through self-consciousness.[111]

Unlike the West, where competition amongst physicists has discouraged discussion of research findings before publication, group loyalty and tight networks in Japan have encouraged more open exchange of ideas. One special characteristic of the successful development of meson theory before and during World War II was that it was done on the basis of collaborative research and discussion. Whether the Japanese are traditionally group-oriented is a matter of much debate. Nevertheless, for the physicists, it served as a useful formula for success. This way of tackling research would be encouraged at the Research Institute for Fundamental Physics, INS, and other joint-use research institutes. But within the group, there were differing influences and inspirations.[112]

Figure 5.5 Hideki Yukawa seated in his home, doing calligraphy, March 1962.

Courtesy, AIP Emilio Segrè Visual Archives, Yukawa Collection.

It seems that Tomonaga's style of research was to do much of the calculations himself and to have his students check or complete them for him. This somewhat individualistic approach may have been influenced by his stay with Heisenberg. Yukawa, on the other hand, tended to air his ideas in front of his colleagues and students, and develop them with their assistance. Nishina, too, liked discussing his ideas. Sometimes, his study meetings began at noon and continued on until 10 P.M. without a break.

Tomonaga and Yukawa led fairly protected lives as children, with little contact with Marxism, which was quite fashionable at the time. As a result, the two had little awareness of social problems facing the Japanese. Both Tomonaga and Yukawa were influenced by Nishida philosophy, which attempted to mesh Eastern and Western philosophies.[113] Such factors help to account for differences between Sakata, Tomonaga, and Yukawa. Such differences in methodological approaches to research and attitudes to the larger society did not make for direct conflict as the three were spread throughout Japan. Yukawa (Kyoto University), Tomonaga (Tokyo University of Education), and Sakata (Nagoya University) formed a triangular relationship, which geographically mapped out Japanese physics.[114]

Unlike Sakata and Taketani, Tomonaga and Yukawa came in late to the social arena. In the aftermath of Nishina's death, Tomonaga was veritably thrown into promoting nuclear physics, high-energy physics, advising on policy and reforms of the research system and university management, and striving for the abolition of nuclear weapons. Such a change also occurred in Yukawa, especially after the receipt of the Nobel Prize.[115] Tomonaga and Yukawa derived their status foremost as experts in physics. This gave them the authority to take up positions of leadership in the Tokyo University of Education and Science Council for Tomonaga, and the JAEC and Research Institute for Fundamental Physics for Yukawa. Their elevation to Nobel laureates reinforced the perception of public "expert," especially in things "atomic" or "nuclear." In light of their newly gained ability to influence the public, other scientists, and government, Tomonaga, Yukawa, and Sakata organized the Kyoto Conference of Scientists to provide a Japanese answer to the Pugwash movement.

Such activities disguised from the public the rivalry and considerable degree of antagonism between Yukawa and Tomonaga. Their opinions frequently clashed and they never produced a jointly authored scientific paper. Yukawa was an aloof, latter-day Confucian-type scholar who was ensconced in his research institute. Whereas Yukawa's manner bordered on the aristocratic, Tomonaga's was like a commoner's.

Tomonaga was a much more approachable person, affable and generous to the many students who flocked to him. Foremost, they recall his sharp mind and lively sense of humor, which he put to good use at parties as a "rakugo" (comic stories) raconteur, seated Japanese style

Figure 5.6 Sin-itirō Tomonaga wearing a kimono and standing outdoors in his garden, ca. August 1970.

Photograph: Hiroshi Chiba. Courtesy, AIP Emilio Segrè Visual Archives.

on his forelegs on a cushion, and often clad in kimono (figure 5.6).
Through such activities, he showed that he was part of the group.
Although having dissimilar personalities, Yukawa and Tomonaga
were, nevertheless, able to cooperate when it came to arms control
and other matters of public concern.[116]

VI   YUKAWA AND TOMONAGA: THE PEACEMAKERS

Japanese physicists took up the discourse of pacifism out of social
responsibility and in the hope of joining the world community. Both
internationalism and pacifism relate to an awareness that one's
existence depends on the cooperation and well being of the rest of the
world.[117] Embracing both showed that physicists supported humani-
tarian values and were worthy of receiving public support and
government funding. While Yukawa and Tomonaga were committed
to favoring basic over applied research, and were more interested in
regulation of science than in its potential commercialization, they felt
that they were especially qualified to deal with the aspect of science
policy dealing with international issues.

In the late 1950s, elite scientists throughout the world were
playing leading roles in the nuclear disarmament movement. But
although Sakata, Tomonaga, and Yukawa were active in the peace
movement in Japan, they were slow to oppose nuclear weapons.
Despite participation by physicists in survey groups to Hiroshima and
Nagasaki, the Japanese did not organize themselves to protest against
the testing of the bomb until the 1950s, after the initiative shown by
the Pugwash movement. This reflects a perception of the atomic
bomb quite different from that held by their American colleagues.
Soon after the end of the war in 1946, Taketani interpreted the bomb
as divine punishment wrought by the world's scientists on Japanese
militarism. This image of the bomb even extended to the Japanese
Communist Party, which initially viewed the Occupation Forces as
liberators of the Japanese. Such perceptions of the bomb and of the
Occupation Forces were nurtured by SCAP, with critical commentary
on the effects of the bomb and any anti-American writing being
censored by them. Moreover, as the Japanese did not make much
progress on their own atomic bomb, the collective guilt of physicists
did not quite extend to Japanese physicists.[118]

A change in Yukawa's thinking may have occurred during his visit
to Princeton where he met Einstein. Einstein emotionally apolog-
ized for the death and destruction brought on the Japanese by the
bomb. But it was only after the Bikini Atoll hydrogen bomb test on

March 1, 1954 that the Japanese movement to ban atomic and hydrogen bombs really got going. Yukawa was one of 12 people who formed a group to collect signatures for a petition to ban atomic and hydrogen bombs. He was also one of the signatories to the inaugural statement of August 8, 1954. This group became the organizing committee for the World Conference Against Atomic and Hydrogen Bombs, which was held at Hiroshima during August 6–8, 1955. Yukawa was also one of 11 signatories to the Russell–Einstein Manifesto, authored by 8 Nobel Prize winners and other scientists.[119]

Yukawa and Tomonaga showed that scientists could be instruments of peace, their social status, rather than know-how, giving them their authority. "Peace" provided the relatively nonpolitical Yukawa and Tomonaga with an issue around which to organize themselves. Campaigning for nuclear disarmament suited them for it avoided the tendency of a Cold War mentality and depiction of one side as good or bad. The dangers posed by both superpowers could be renounced without taking an anti-American or anti-Soviet stance. What's more, they considered themselves as being eminently qualified for such a role. Being Japanese (the "victim" of the bomb) and physicists (the creators) they knew all about nuclear devastation. Pacifism was, in fact, a means by which the hitherto apolitical Yukawa and Tomonaga became figureheads to an essentially political type of activity. Such concern of scientists for liberal causes while initially nonpolitical, nevertheless, served to draw them into socialist politics.

In 1955, Yukawa became a member of the World Peace Appeal Group of Seven Committee, along with Saburō Shimonakaya, Seiji Kaya (professor in metallurgical physics, Tokyo University), and others. They were later joined by Tomonaga and the Nobel Prize–winning novelist Yasunari Kawabata.[120] The group served as part of the League for the Establishment of a World Federation of Nations. Yukawa's stance was a rather idealistic one when compared to Tomonaga's who tended to be more aware of political realities. But both played the role of statesman with little contact with grassroots movements. Unlike Tomonaga and Yukawa, Sakata was more concerned with taking action and was closely involved with the scientist movement. The Pugwash Conferences on Science and World Affairs were a manifestation of the desire of scientists to remain as public watchdogs. In 1957, the First Pugwash Conference was held in Canada. Many of the participants were physicists. Yukawa, Tomonaga, and Yukawa's nephew Iwao Ogawa, who had conducted research on radiation, were amongst the 22 participants. Tomonaga wrote about the conference in an article entitled "Kagakusha toshite hatsugen suru" ("Speaking as a Scientist") in the magazine *Sekai* (*The World*) (September 1957). This was the first major public comment on a social issue by Tomonaga, a physicist who had previously been reticent to become involved in such matters.[121]

The Third Pugwash Conference was held in Austria. Yasuo Miyake and Shōichi Sakata joined the original three Japanese participants there. In 1958, Yukawa became an adviser to the World Association of World Federalists (WAWF). At the 1961 world conference in Vienna, he was elected president, continuing in that position till 1965.

In 1962, the first meeting of the Kyoto Conference of Scientists was held with Yukawa, Tomonaga, and Sakata being the key organizers. The three were assisted principally by physicists: Toshiyuki Toyoda, Iwao Ogawa, Shūji Takagi, Eiji Yamada, Zirō Maki, and others. The concept of nuclear deterrence was strongly criticized at this conference.[122] Yukawa hoped to develop a strategy through the Kyoto Conference that was much more in keeping with the original Russell–Einstein Manifesto, which called for the abolition of nuclear weapons. Another feature of the Kyoto Conference was that it included participation by social scientists in addition to practitioners of the natural sciences. Japanese participation in both tended to center on Sakata, Tomonaga, and Yukawa. Yukawa and Tomonaga, were more interested in using the Kyoto Conference as an elite discussion/study group rather than as the core of a mass scientist movement. This suited Yukawa in particular for he had always been a fairly reclusive figure; internationalism and mass movements were not really his style.[123]

Yukawa's and Tomonaga's approaches were similar to that of the Russell–Einstein Manifesto in that they were utterly opposed to nuclear war and criticized the idea of nuclear deterrence. For them, the political activity of such pronouncements was quite separate from their research, whereas to Sakata and Taketani, they were strongly linked.[124] Sakata, however, was concerned with the broader scientific community in what Nakayama describes as Bernal–Sakata-ism: a happy marriage of scientism and Marxism.[125]

The enthusiasm for international peace continued with Yukawa presiding over the World Congress of the WAWF, which was held in 1963 in Tokyo and Kyoto. Yukawa called to other countries to follow Japan's example of a "peace constitution" and renounce war. He had hopes for a revitalized United Nations, which would become a true world federation. He felt that world federation should, however, be achieved concurrently with a global reduction in armaments.[126]

The second Kyoto Conference of Scientists was held in May 1963 at Takehara in Hiroshima prefecture. The conference stressed on how

Japan should adhere to a policy of non-nuclear armaments and continue to (officially at least) refuse to allow the entry of nuclear weapons into Japan. The third was during the period June 30–July 2, 1966, in Tokyo. The meeting criticized the idea of a "balance of power." Yukawa and Tomonaga (minus Sakata who had passed away) helped to organize the Pugwash International Symposium on "A New Design Towards Complete Nuclear Disarmament: The Social Function of Scientists and Engineers," which was held in Kyoto from August 28 to September 1, 1975 and, during which Yukawa and Tomonaga issued a statement entitled "Beyond Nuclear Deterrence." Because of student unrest from around 1968 and the death of Sakata in 1970, the fourth Kyoto Conference of Scientists was not held in Kyoto until June 1981. The fourth conference once again reiterated the previous positions of opposing nuclear deterrence and a nuclear balance of power, and provided further calls for maintaining Japan's three non-nuclear principles, which had been adopted by Japan's parliament in 1971.[127] Under these principles, Japan rejected both the possession and the production of fissionable weapons, and opposed the allowing of nuclear warheads to be brought into the country. The conference was held almost as a final farewell to Yukawa and the type of physicist–public man that he represented. Shortly after, on September 8, he passed away.[128]

The conferences that Yukawa, Tomonaga, and Sakata led, sought to influence the public and decision-makers through the media rather than through grassroots movements involving the Japanese public. Endowed with a special authority by courtesy of their being called physicists, they constituted an elite within an elite, which they hoped would save the world from the bomb.[129] While the effectiveness of the scientists' appeals for the abolition of nuclear weapons on governments and the general public is difficult to evaluate, the world scientific community provided a channel for dialogue that transcended Cold War divisions. The willingness of Japanese physicists to visit the People's Republic of China and the USSR may have irritated the United States and the Japanese Ministry of Foreign Affairs, but it did serve to set up interaction between the countries, which was otherwise not encouraged.[130]

CONCLUSION

Physicists portrayed themselves as a bulwark against the profit-hungry private sector and military-minded politicians. For example, in order to placate the Tanashi townspeople prior to the establishment of the

INS, Tomonaga and his colleagues were put into the position of committing themselves as adversaries to applied atomic energy research and any research of a military nature. This self-polarization as a safeguard against the self-interest of the private sector meant that physicists who did not conform to this image were exiled to the academic periphery. Physicists who embraced entrepreneurial or technocratic values were not considered "public men" for they would be corrupted by business and the pursuit of wealth. Physicists who did become involved in private industry lent their stamp of validity to the activities that they collaborated in.[131]

The propensity of Tomonaga and Yukawa to become public men can be traced to the public-spiritedness of the samurai who were their ancestors and the notions of public service that were implicit in the minds of their parents. Sakata, some four years their junior, was also acclaimed for his research in physics. Although reputedly nominated for the Nobel Prize, he failed to be awarded the ultimate validation as an expert and senior statesman of science. Sakata preferred hands-on involvement in social issues, whereas Tomonaga and Yukawa chose to somewhat distance themselves from the populace, as might befit senior statesmen. All three sought to influence policy for science, but they also felt that science could guide public policy and help achieve world peace. The importance of Yukawa and Tomonaga as experts lay in the fact that they were relatively apolitical and highly esteemed by the people in their profession. To the Japanese, Tomonaga's Nobel Prize in 1965 was late in coming, but it served to validate what they had always known—that he was an expert.

CHAPTER 6

CORPORATE SCIENCE: SAGANE

When the atomic bomb was dropped on Nagasaki on August 9, 1945, an anonymous letter addressed to Ryōkichi Sagane was attached to three recording instruments that were parachuted by the Americans.[1] The handwritten letter implored Sagane

> to do your utmost to stop the destruction and waste of life which can only result in the total annihilation of all your cities, if continued. As scientists, we deplore the use to which a beautiful discovery has been put, but we can assure you that unless Japan surrenders at once, this rain of atomic bombs will increase many fold in fury.[2]

The letter was recovered around 3 P.M. on August 9, some 50 km from where the bomb was dropped. The "three colleagues" who had sent the communication were Luis W. Alvarez, Robert Serber, and Philip Morrison, three physicists whom Sagane got to know while working with Ernest Lawrence at Berkeley. The three had later participated in the Manhattan Project and were on the strategic island of Tinian, where parts of the bomb had been stored, at the time the bomb was dropped.[3] When visiting Berkeley in 1949, the local press exaggerated what had happened.

> Prof. Ryokichi Sagane, of Tokyo University, related how a note, addressed to him, was dropped in a transmitting tube during the Nagasaki bombing. . . . it fell into Japanese Naval Intelligence hands. However, the military finally sent for Sagane for "possible reassurance" about the deadliness of the bomb. "I just told them there would be a third attack if they didn't give in," the Japanese said. "The Americans aren't fooling with this bomb, I said." The next day Japan capitulated to the U.S.[4]

In September and October 1945, the U.S.-led Scientific Intelligence Survey interviewed Sagane and his father, Hantarō Nagaoka, at the large, but run-down, Tokyo University physics lab. A large Van de Graaff generator (Sagane's handiwork) caught the eyes of the interviewers (Drs Karl T. Compton and David T. Griggs) sent to question Sagane and Nagaoka. They also were impressed by an electromagnetic spectrograph, which would allow particles of different properties emitted from a neutron-bombarded target to be collected.[5] The interviewers considered Sagane's talents wasted during the war.[6]

Immediately after the end of the war, Ryōkichi Sagane in Tokyo wrote to Ernest Lawrence in Berkeley describing the daunting work that had to be done to rebuild Japan:

> these days I am still busy from several different kinds of tasks urged to be fixed promptly. Firstly the problem of reorganization of science, secondly presentation of opinions about or results of . . . the problem of reformation of universities, thirdly devising direct practical measures for the new education system, fourthly help for the industrial production in Japan by supplying technical advices or solving urgent problems for the purpose, and research works on high vacuum and on problems related to the Van de Graff generator comes to the last.[7]

Despite the imperfect English, what is clear is that Sagane felt that the basic research, which he had been involved with before and during the war, was least important, but not off the agenda. This would be the story of Sagane's postwar career, his values changed by the harsh economic realities that Japan faced at the time.

So far we have concentrated on the activities of university-based physicists and one politician. In chapter 4, we traced Sakata and Taketani's fight for social change and social responsibility in science, through democratization of the research system, policy initiatives via the Science Council, and prolific writings. Chapter 5 took us closer to the actual governing elite. The Nobel laureates Yukawa and Tomonaga attempted to influence the policy-making process from the vantage point of highly respected university professors. Yukawa rubbed shoulders with politicians, bureaucrats, and corporate leaders as a member of the Japan Atomic Energy Commission (JAEC), but soon retreated to the confines of his own institute at Kyoto University to engage in research and pacifist activities. Tomonaga had a long involvement with the Science Council, especially as president for much of the 1960s. He was active in research infrastructure building, helping to establish the Institute for Nuclear Study as a model inter-university facility run along democratic lines. Yukawa and Tomonaga refrained from overtly political activities, choosing a platform of "peace" when they wanted to bring about change. They were quiet achievers. But the civilian nuclear program required qualified personnel with technical expertise to transfer technology and to train management. Ryōkichi Sagane was one of those people—a physicist who left academia and became immersed within the technocracy. He, like Nishina, saw that science policy needed to be not only concerned with the promotion of research, regulating the use of it, and international issues. The commercialization of research into new products or services was critical if Japan were to prosper.

Why the career change? Sagane's privileged background and career in experimental physics, to be outlined in the first section, show that he, of all the physicists, had the technical experience to pursue the development of civilian nuclear power. With World War II and Japan's defeat, Sagane joined the "brain drain" of top physicists to the United States in search of a better research environment. In the second section, we will see how Japan's fledgling atomic energy program provided him with a reason to return in the mid-1950s. He provided a human link between Japanese politicians pushing for atomic energy and top physicists and companies in the United States and the United Kingdom. His role in the establishment of an infrastructure for civilian nuclear power is outlined in the fourth section. Sagane's willingness to introduce foreign technology is contrasted with his view, given in the fifth section, that nuclear fusion should be pursued by first building a base in basic research overseen by the Ministry of Education. In sum, this chapter suggests that physicists did make a difference in science policy, and that even as a government-employed expert, an "insider," Sagane was able to maintain an independent voice.

I THE GETTING OF WISDOM

Unlike many other government scientists or administrators, Sagane did not enter civil service as a new university graduate. He occupied the plum post of professor at the prestigious Tokyo University and spent a total of eight years overseas. The chance of working on the development of atomic energy in the 1950s was one that he could not refuse. Friends in both the United States and Japan encouraged him to pursue this promising area.

Sagane's biography suggests that socially and institutionally, he was well placed to influence the direction of Japan's civilian nuclear power program. Born in Tokyo on November 27, 1905, he was the fourth son of the eminent physicist Hantarō Nagaoka. Many of Ryōkichi's brothers found employment in engineering-related areas. He, too, was interested in technology, but after encouragement from his father, decided to pursue a career in physics.[8] In 1914, he was adopted by Mrs. Chiyo Sagane, thenceforth taking on her family name.[9] His education followed the prestigious academic route—graduating from the First Middle School in 1923, the First Higher School in 1926, and the Department of Physics, Tokyo Imperial University, in 1929 where he would later become professor. Like Yukawa and Tomonaga, there was a definite pattern of occupational ascendancy in academic families who, more often than not, were well-to-do.

Sagane entered graduate school in April 1929, studying under Kiyoshi Kinoshita and Takeo Shimizu, the latter being particularly interested in radioactivity. He finished postgraduate studies in June 1931 when he took on the status of research student at Riken. The following month he became one of the first members of Nishina's lab when Nishina became a chief researcher. Sagane set about making Geiger counters for use in cosmic ray research. This activity was taken over by Masa Takeuchi, another early lab member, and Sagane turned his talents to the design of a cloud chamber.[10] Sagane would go on to become a central member of Nishina's group, with involvement in most of the experiments carried out. In April 1933, he was appointed lecturer in the Faculty of Science at Tokyo Imperial University, but, nevertheless, remained busy at the Nishina lab. In 1934, for example, he participated in experiments with four groups, investigating nuclear transformations, cosmic rays, neutron research, and positron studies, respectively.[11]

From August 1935 to November 1936, Sagane conducted research at Lawrence's Radiation Laboratory, after which he visited the Cavendish Laboratory, Cambridge, and other research institutes in Europe. Sagane arrived in San Francisco on September 4, 1935 to study at Lawrence's Radiation Laboratory, for a stay of over a year, at about the same time as Luis W. Alvarez.[12] Both began performing very menial tasks. During the first two months, these included looking for vacuum leaks in the cyclotron chamber and lacquering of suspected leaking joints. Eventually, both were given cyclotron time to look for new radioactivity.[13] In March 1936, during his absence, Sagane was promoted to associate professor at Tokyo. After visiting Europe, he returned to Berkeley in June 1937, staying there till February 1938.[14]

Construction of a 26-inch cyclotron commenced at Riken and was completed by April 1937. The Nishina lab benefited from advice given by Sagane while he was abroad. Sagane showed more interest in techniques than in physics for its own sake. Although Sagane thoroughly enjoyed his stay in the United States and would have preferred to have continued to live in California, the worsening relationship between the United States and Japan meant that he had little alternative but to return to Japan. From January 1939, he was employed part-time at Riken, receiving his DSc in December 1939 with a dissertation entitled "On Artificial Radioactivity."[15]

Around the beginning of the Pacific War, Sagane was often called to help in military research and in May 1942, was made a full research member of Riken, acknowledging his central role in nuclear physics experiments and in the construction of the prewar cyclotron. In March 1943, he was promoted to professor at Tokyo Imperial University. That August, he joined the Naval Technical Research Institute to work on vacuum technology as there was a great shortage of reliable vacuum tubes in Japan.[16] He was involved in policy-making from 1943 when he became a member of the National Research Council. In January 1944, he became a special consultant to the Army, working on a similar project at the Tama Research Institute. Sagane made contact with companies such as Japan Radio and helped complete magnetrons at the Shimada Naval Technical Research Institute, along with Tomonaga, Masao Kotani, and others. Unlike Tomonaga's wartime work, Sagane's efforts did not result in any major achievements. In June 1944, in the middle of the mobilization of science, he was placed in charge of the National Research Council's No. 138 Research Group.[17] In August 1945, soon after the dropping of the atomic bomb on Hiroshima, Ryōkichi Sagane joined a survey group to examine the damage wreaked on the city.[18]

When interviewed by Compton and Griggs in late 1945, Sagane was much concerned as to whether or not he would be permitted to continue research in nuclear physics, but was advised that provided it was not of a military nature and did not involve the concentration of a large amount of unstable radioactive materials, there should not be a problem. GHQ was to be kept informed of projects.[19] Sagane did his best to rebuild the lab at Tokyo University by reviving the Van de Graaff generator and encouraging his students to return to research.

Harry Kelly, chief of the Fundamental Research Branch of the Scientific and Technical Division, Economic and Scientific Section (ESS), Supreme Commander for the Allied Powers (SCAP), formed the Scientific Liaison Committee with three members of the Japanese scientific community, all based at Tokyo University: Sagane, Seiji

Kaya, and the botanist Hiroshi Tamiya. Sagane's ability to speak some English enabled him to make representations on behalf of his fellow scientists, and was crucial in convincing Kelly of the need to establish a genetics research institute at Mishima. Furthermore, it appears that Sagane was able to even comprehend the Australian accent of Brigadier John O'Brien, the chief of the Scientific and Technical Division of ESS.[20] The Scientific Liaison Committee paved the way for the establishment of the Japan Association for Science Liaison (JASL) and later the Renewal Committee for the Organization of Science of which he was also a member (August 1948–January 1949).

Sagane had always shown an interest in engineering. At a number of JASL-related meetings from late October and November 1946, during discussions regarding the reorganization of science, Sagane discussed with Kelly and others the idea of establishing a separate group for liaison with the engineering field. This resulted in the formation of an engineering liaison group in December. This would, in turn, become the Japan Association of Engineering Liaison.[21]

In 1946, the Atomic Energy Act was passed by the U.S. Congress. The Act, in effect, created the Atomic Energy Commission (AEC) and placed severe restrictions on the dissemination of nuclear research. The use of radioisotopes in Japan can perhaps be considered the first instance of the postwar peaceful use of atomic energy. This was, at least in the eyes of the Americans, put in a separate category to nuclear power generation. The scientists and the Japanese Government saw the need for a national program of nuclear research but the Far Eastern Commission still thought it best, in January 1947, to maintain the prohibition on all research in Japan in the field of atomic energy.[22] In 1949, the U.S. AEC approved the use of radioisotopes in Japan by Yoshio Nishina and what had formerly been Riken. The radioisotopes were supplied the following year. This approval was to a large extent motivated by Nishina's publicization of the destruction of the cyclotrons by the Occupation Forces and the resulting outrage of American physicists such as Ernest Lawrence.[23]

Sagane, too, resolved to utilize his most important overseas contact. Writing to Lawrence of his desire to visit the United States, Sagane suggested that

> Thinking . . . what I should do in USA, it is not the proper research work for myself but [*sic*] first comes in my mind that I have to make effort to send back many informations to Japanese scientists about the recent progress made in the past few years in the world. Thus give them their earliest start to work in their full activity.[24]

Sagane thus envisaged for himself a role as a "bridge" between the United States and Japan for other Japanese. Kelly facilitated this by writing to Professor Gerald W. Fox, mentioning that

One of the men who is very eager to go back for even so short a stay as two months is your old friend Dr. Ryokichi Sagane. Sagane, as you know, is one of the most competent experimental scientists here, and I am surprised that nobody has as yet extended an invitation to him. Would it be possible for Iowa State College to extend such an invitation?[25]

Fox agreed. In December 1949, Sagane was given permission to leave the country to accept a visiting professorship in the Department of Physics, Iowa State College, from January 1 to July 1, 1950.[26] On his way there, he stopped at Berkeley and met Alvarez again.[27]

Since Sagane had last visited Berkeley, much had changed. In May 1947, C.F. Powell and his colleagues at Bristol University confirmed the existence of two types of mesons in cosmic rays—the heavier mesons (pi-meson or pions) interact strongly with nucleons (protons and neutrons) and decay into the lighter mesons (mu-mesons or muons). Lawrence invited Sagane to visit the Radiation Laboratory, while he was in the United States, as his guest to work on pion physics. A 184-inch cyclotron at Berkeley had started to produce the first artificially made pions. Sagane invented a spiral orbit magnetic spectrometer, which Lawrence had made up for muon experiments. The end-result became known as the "Sagane Magnet."[28]

Sagane was part of the brain drain of outstanding physicists such as Yukawa, Tomonaga, Seishi Kikuchi, Satio Hayakawa, and Yōichirō Nambu who had gone to the United States. Lawrence drew on a "small private fund" to support visits by Sagane and Kikuchi. He also approached the Ford Foundation in the hope that Japanese scientists might participate in its Exchange of Persons Program, describing such visits as "an extraordinary opportunity" to promote U.S.–Japan relations.[29]

Sagane certainly found the research environment to his liking for he extended his period overseas till April 30, 1955, a period of over five years. This was an unusually long period away for an academic with a university position in Japan, but it did serve a useful purpose for colleagues visiting the United States, and Sagane made a good ambassador for Japanese physics. He was likely to have been instrumental in Ernest Lawrence's visit to Japan in 1951. On February 28, 1955, Sagane resigned from Tokyo University because of his protracted stay in the United States. Out of a job, the prospect of the development of atomic energy in Japan suggested exciting work opportunities.

II ATOMS FOR PEACE

The prohibitions placed on the development of aerospace technology and nuclear power by the Allied Powers immediately after the end of the war effectively provided the United States with a ready customer in both of these areas, for the prohibitions on research related to atomic energy did not prevent growing interest on the part of physicists and industrialists in the scientific and commercial possibilities offered by this new energy source. The editorial of the *Asahi shinbun* of February 3, 1948, expressed hope that once an international system for the control of atomic energy had been put into place, "all means of peaceful services will be opened to mankind," and concluded with the line that "We are anxious to have such a day come soon."[30] Kōji Fushimi, professor at Osaka University and one of the enthusiastic scientist–proponents of nuclear power, argued at a 1951 general assembly of the Science Council of Japan (JSC), that the soon-to-be signed peace treaty should not contain any clauses prohibiting nuclear energy.[31] American companies hoped so too.

The Allied Occupation helped to form ties between American industry and Japanese utilities and equipment manufacturers, often with militaristic overtones. The decision of the U.S. AEC, on November 11, 1949, to allow Japanese participation in the program for the distribution of radioisotopes came shortly after the communists had taken power in China. This was despite prohibitions on experiments relating to atomic energy. Technological ties were substantially strengthened during the Korean War, when Japan imported American industrial technology under the auspices of a special procurement program. In 1952, the government considered the establishment of a science and technology agency for purposes of national defense, somewhat reminiscent of pre–World War II mobilization.[32] Given the military imperatives that seemed to have been behind Japan's concern with science and technology in the past, and even after the war, how would the Japanese concern with "peace and democracy" affect its propensity to mobilize science and technology in the future?

The concept of "atoms for peace" provided the Japanese and the rest of the world with an ideal of a prosperous future courtesy of science. The Japanese were able to reconcile the establishment of a nuclear industry in their own country with the devastation that had been wrought in Hiroshima and Nagasaki less than a decade earlier. An atom that had been "tamed" would work for the benefit of "mankind." Sagane, whose name was virtually on the Nagasaki bomb, was convinced of this and played a major role in harnessing atomic energy to serve Japan's "peaceful" power needs. It would only be much later that the Japanese would see that there is often no boundary between technology for civilian and military purposes, and come to the realization that the "sunny side" of the atom was perhaps wishful thinking.[33]

On December 8, 1953, President Eisenhower presented his plan of "Atoms for Peace" to the General Assembly of the United Nations, which set the stage for international cooperation in atomic energy and the creation of the International Atomic Energy Agency (IAEA). In January 1954, one month after Eisenhower's statement, Ichirō Ishikawa, the president of the Federation of Economic Organizations (FEO or "Keidanren" in Japanese), made an official visit to the United States on the inaugural trans-Pacific service of Japan Airlines. He visited San Francisco where he contacted Sagane, who suggested he visit the Lawrence Laboratory at the University of California, Berkeley.[34] In September 1954, an AEC member Thomas E. Murray urged the United States to build a nuclear power plant in Japan to show the world (and especially the Soviet Union) how benevolent the United States was.[35] Internationalist ideals were combined with nationalist policies aimed at enhancing American scientific prestige.

In the initial stages of interest in atomic energy, those in government and industry first turned to technocratic physicists such as Sagane, Yoshio Fujioka, and Kōji Fushimi for advice. Seiji Kaya, professor at Tokyo University, and Fushimi both felt that an AEC was needed in Japan. Kaya voiced his opinion that there was a need to pursue nuclear power at an April 1952 meeting of Division Four (Fundamental Science) of the Science Council. One of the first public pronouncements regarding the need to pursue atomic energy research was uttered by Kaya at the opening of Yukawa Hall, also known as the Research Institute for Fundamental Physics, at Kyoto University on July 20, 1952. He suggested that a fact-finding mission be sent overseas and that an AEC be formed in Japan, which would be responsible to the government. Kaya felt that it would be difficult to establish such a body within the framework of the JSC. Fushimi also felt that such a commission was needed.[36]

In late 1952, at the thirteenth general assembly of the JSC, scholars rejected a proposal by Seiji Kaya and Kōji Fushimi to set up such a body. Instead, it was recommended that Council members survey the opinions of their respective associations regarding nuclear power and that a special committee be established to prepare a proposal to be presented to the general assembly in April 1953.[37] A committee was formed to advise on this: the Thirty-Ninth (Temporary) Committee.

Meanwhile, overseas developments included the establishment of the UK Atomic Energy Authority (AEA), and approval of plans for the construction of the world's first large-scale nuclear power reactor: a 138-MWe (megawatts electricity) gas-cooled graphite-moderated Calder Hall reactor. (By the end of 1956, the facility was successfully generating electricity.)[38]

Some physicists were cautious and suggested that atomic energy policy be based on the results of sound scientific research carried out in Japan, whereas the industrialists who became involved tended to see it as another moneymaking venture. This was one of the fundamental reasons for the long conflict between scientists critical of nuclear power and the government–industrial complex, which included their more technocratically minded scientist–colleagues. The former was committed to the importance of a domestic research base whereas economic and industrial activities in Japan from the 1950s tended to be based on imported technology. The physicists, consequently, found themselves as advisers to the government on questions of nuclear safety rather than on the actual technology itself.

III   NAKASONE AND SAGANE

As early as 1951, Nakasone had visited the physicist Seiji Kaya, making enquiries about nuclear power. He went to the trouble of going to the United States around 1953 and speaking with Sagane who was at the Lawrence Laboratory at Berkeley at the time. It appears that this was prior to Nakasone attending a summer school at MIT on atomic energy. Nakasone apparently emphasized to Sagane how important nuclear power was to Japan and how he had hopes of establishing a Japanese program.[39] In March 1954, after returning from the trip abroad, Nakasone made a budget request for an additional 250 million yen for science and technology, 235 million of which would be allocated for the construction of a nuclear reactor and 15 million toward uranium prospecting.[40] Nakasone was the main promoter of the surprise budget request in the minority political party called "Kaishintō" ("Progressive Party"). The party had the casting vote in terms of passing the 1954 budget, and was, therefore, successful in having the proposal supported by the other two conservative parties, the Liberal Party and the Japan Democratic Party.[41] This was approved in great haste by the House of Representatives, against the wishes of the Science Council. The Council would have preferred that atomic energy policy first be debated before any budget allocation was made.[42] That initial atomic energy budget set the tone for the estranged relationship between government–industry circles and their scientist–critics.

Seiji Kaya and Yoshio Fujioka visited Nakasone to discuss the budget submission. Kaya's recollection of Nakasone's words is as follows:

> Originally such kind of research budget should be submitted to the Diet through the Government by the requirement of researchers, or by the Diet members under the consultation with the researchers concerned. The former case was not able to adopt [*sic*], because the attitude of Science Council, which is the representative organ of Japanese scientists, was negative to initiate the study of atomic energy. For taking the latter attitude, unfortunately we had no time to make consultation with scientists. However, atomic energy is now estimated as the most important energy source in the future. Considering the brilliant progress made recently in USA, UK, France, Soviet Union, etc., in the peaceful use of atomic energy, we, Diet members, can not help to appropriate some amount of research fund for the study of atomic energy even in such abnormal way [*sic*] for the purpose of discharging the heavy responsibility of the Diet members.[43]

The bulk of the atomic energy budget was administered by the influential Ministry of International Trade and Industry through a committee that it organized. The Government also set up a Preparatory Committee for the Peaceful Use of Atomic Energy, which would lead to the establishment of the JAEC.[44]

The year 1954 was a good one for Nakasone. In August 1954, Nakasone visited Northern Europe, USSR, and the People's Republic of China to facilitate the repatriation of Japanese internees and war criminals, and the return of Sakhalin and the Kurile Islands. Ichirō Hatoyama became president of the new Japan Democratic Party and Nakasone became head of the Japan Democratic Party Organization Bureau. Nakasone was reelected, for the fifth time, along with the Ichirō Hatoyama government.[45] Hatoyama took over the prime ministership from Yoshida, and sought to revise some of the reforms that were implemented during the Occupation period. He promoted policies aimed at a stronger nation through strengthening of the police, and greater central control of education. He unsuccessfully attempted to revise Article 9 of the Constitution, which prevented Japan's rearmament.[46] These nationalistic policies were much closer to Nakasone's own political agenda than that of Yoshida or the later prime minister Eisaku Satō.[47]

Meanwhile, scientists were becoming concerned. It was felt that a permanent Japan Science Council committee should be established to deal with problems relating to atomic energy. The Committee for Atomic Energy Problems was created to take over from the Thirty-Ninth Committee, and its first meeting was held in May 1954. The committee consisted of 14 members with the first chairman being Shōichi Sakata.[48] In 1954, the JSC Special Committee for Nuclear Research (SCNR), headed by Sin-itirō Tomonaga, took on the responsibility of deliberating upon an appropriate policy for nuclear power. Taketani's ideas provided the SCNR with a base upon which to build on.[49]

On March 16, 1954, the *Yomiuri shinbun* newspaper reported that 23 Japanese on a tuna fishing boat called "The Lucky Dragon" had been contaminated by radioactive fallout near the Bikini Atoll during hydrogen bomb testing about 2 weeks earlier. This had inauspiciously occurred on almost the same day that it was decided to present Japan's first reactor budget before parliament. The incident served to pour cold water on the attempts to begin a Japanese nuclear power program.[50] Nevertheless, on June 9, the FEO made a submission to the government urging the establishment of a science and technology agency, this time not for defense purposes but for the "promotion" of science and technology, and to help formulate and administer government policy, especially nuclear power.[51]

The JSC was often in conflict with the Government over the importance of basic research. The nuclear physicist critics in the JSC expressed their displeasure at the large budget appropriation for atomic energy, particularly since it had been done in the absence of a properly formulated policy and without consultation. At the autumn JSC meeting, a statement was made that lay down three basic principles or conditions necessary for nuclear research: (1) non-secrecy; (2) democratic control; and (3) autonomy. Despite the concern of some scientists for the peaceful use of nuclear power, the close collaboration between government and industry effectively excluded much input from the more critical-minded scientists.

The Japanese rush to establish an atomic energy program was partly a result of a desire to take full advantage of the possibilities of transfer of nuclear technology from the United States under the Atoms for Peace plan. The beginning of the program was, like many of Japan's other postwar industries, characterized by a great reliance on scientific and technical information from abroad. This was reflected in the early strategy of Japan's nuclear program, which was, simply, to have a commercial facility in operation as quickly as possible. This was most easily achieved by importing the appropriate technology.

In 1954 and 1955, there was considerable support and interest in atomic energy across the political spectrum by politicians, bureaucrats, business leaders, and academics. In May 1954, Cabinet established the Preparatory Council for the Peaceful Uses of Atomic Energy. It was hoped that atomic energy would eventually provide the Japanese with an independent energy source. Scientists argued the need to conduct basic research in atomic energy. The concept of energy "self-sufficiency" was used by many proponents of nuclear power. But given that the United States was relied upon to provide much nuclear technology, and since the United States controlled the uranium supply, autarky was only achieved in so far as it assumed a continuing U.S.–Japan relationship.[52]

Sagane facilitated the growing technological ties between the United States and Japan. Americans had been eager to include Japan in the Atoms for Peace nuclear power program. Thomas E. Murray (US AEC) and Senator Estes Kefauver encouraged the building of power plants in Asia, especially Japan, because it would show the Soviet Union how good America's intentions were. In January 1955, the United States offered Japan access to 20 percent enriched uranium and within months, atomic energy specialists from General Dynamics, Chase Manhattan Bank, and nuclear fuel companies were visiting Japan. On November 14, 1955, the U.S.–Japan Atomic Energy Agreement was signed, only a little more than a decade after the atomic bomb was dropped on Japan and three years after the end of the Occupation.[53] The agreement enabled Japan to buy nuclear fuel and purchase nuclear technology.[54]

In May 1955, the FEO set up a permanent Committee for the Peaceful Use of Atomic Energy, chaired by Daigorō Yasukawa. Fushimi continued to push for the peaceful use of atomic energy, stating his case in the June issue of the periodical *Chūō kōron*.[55] The enthusiasm of the government and industry for nuclear energy was boosted when the First United Nations International Conference on the Peaceful Uses of Atomic Energy was held in Geneva in August 1955, a meeting that showed to the world the wonders of nuclear technology. (The Geneva Conferences of 1955 and 1958 are generally considered to have opened up the subjects of nuclear fission and fusion for energy purposes.) Japanese participants included scientists and representatives from business and government. It was against this backdrop that Seiji Kaya encouraged Sagane to return to Japan and become a key member of the development of atomic energy in Japan. Sagane returned around February 1956.[56]

IV   ESTABLISHING AN INFRASTRUCTURE FOR CIVILIAN NUCLEAR POWER

Some scientists sided with the government and devoted their attention to research and development of civilian nuclear power. This produced a longstanding schism amongst Japanese scientists. Although Yukawa became one of the founding members of the AEC of Japan, he did so in the hope that basic research in theoretical and experimental nuclear physics might benefit from more funding. Yukawa resigned after little over a year, protesting that the political concerns of those in government often took priority over the scientific advice given by certain members of the commission. His highly publicized resignation marked the beginning of a period in which he and other physicists involved in the Science Council went off on a tangent of their own. They, subsequently, set about organizing a long-term plan for nuclear research under the auspices of the JSC, in collaboration with the Ministry of Education, which was responsible for basic science at universities.

The development of atomic energy called for the mobilization of many experts. Sagane, with his ability to negotiate at an international level, experience in technical matters, as well as acquiescence in generally toeing the official line, provided a welcome substitute for more difficult experts such as Hideki Yukawa. His visit to the United States was repeatedly extended till late 1955.[57] Upon his return from the United States Sagane chose to lend his expertise, scientific judgment, and leadership to the industrial world. He facilitated the actual implementation of policy from negotiating the importation of reactor technology to construction of power plants and ensuring their successful operation. He was an insider who influenced the execution of policy, rather so much as a person who determined policies to be adopted. His activities enabled the interests of his "patrons" to be satisfied. Although a highly competent scientist, he was not a particularly distinguished physicist, failing to make highly significant discoveries.

Sagane's identity as a public man, and the form that it would take, was clearly influenced by the economic changes that Japan was undergoing. Lawrence's and Sagane's Japanese friends encouraged him to become involved in Japan's atomic energy program, but leaving academia was a difficult decision. It was a big step to move from one culture to another. This is reflected in a letter from Sagane to Lawrence in late 1956:

> As you mentioned to me before I leave [*sic*] Berkeley, I was persuaded to work in the Atomic Energy Field as soon as I returned to Tokyo.

However, the more I looked into the situation the more I realized the complicated situation in Japan, so that it took me a full half year of hard work before I could make up my mind. Since this summer, I have been involved in the activity of the Japan Atomic Energy Research Institute, as a member of [the] board of directors.[58]

Sagane asked Lawrence's advice regarding the wisdom of having Dr. Lawrence R. Hafstad, head of the Atomic Energy Department of Chase National Bank, visit Japan again to advise on atomic energy for several weeks. The FEO had previously invited Hafstad to speak to a group of business leaders in May 1955 regarding the problems and possible uses of atomic energy.[59] Hafstad was a physicist who had been director of Reactor Development for the AEC since 1949. He would later become vice-president of General Motors Corporation, in charge of corporate research laboratories.[60] Hafstad chided American industry for not embracing nuclear power more quickly.[61] His pro-industry stance was off-putting to many in Japan. Sagane was asked by Matsutarō Shōriki to recommend other American experts who might be appropriate.[62] It was reported that

At his suggestion, the Japanese government invited experts in nuclear power to come over and advise them but due to the influence of the left wing part of the government, they were unable to invite Hafstad. But, they invited representatives from England and France. No one from this country. [USA] They have not approached the Russions [sic]; nor have the Russians approached them.[63]

The choice of experts who were consulted was influenced not only by local politics but Cold War politics as well.

In 1956, Shōriki pushed for the importation of a British Calder Hall-type nuclear power plant. Ichirō Ichikawa, JAEC chairman, was chosen the leader of a group to visit England. Sagane was asked to join the group and ended up becoming a core member. The delegation visited various manufacturing companies and atomic energy facilities in England. Sagane was one of the most active members, and his English, although far from perfect, was of use to the delegation who were mostly unable to speak it. This was at a time when the Japanese still had little experience of international negotiations. Some members of the group had never been overseas, let alone inspected nuclear power facilities or discussed technical matters with foreign experts.[64] During this trip, Sagane became convinced that a commercial power reactor had to be introduced quickly into Japan if utility companies were to become enthused about nuclear power.[65]

Sagane was a pragmatist. Although he understood the desirability of independent development of nuclear power, the technology gap between the United States and Japan was too large. Foreign reactor technology had to be imported and manufactured under license if Japan was to bridge the gap. Many physicists thought otherwise, but Sagane considered their views idealistic, especially given that Sakata, Taketani, Yukawa, and Tomonaga were all theoretical physicists lacking any hands-on experience in building apparatus.[66] Sagane, an experimental physicist, had more in common with his former boss, Nishina, a physicist who had originally been trained as an engineer.

There was a considerable struggle inside the government–industrial complex over who would control nuclear power. The establishment of the Japan Atomic Power Company (JAPCO), to assist in the introduction of a British gas–graphite reactor, represented the compromise that was reached. The company was formed as a joint concern by government and industry, funded 40 percent by private utilities, 20 percent by government utilities, and the remainder by private sources.[67] The government's policy tended to encourage JAPCO and other electric power companies to build nuclear power stations out of their own funds with what were known to be proven types of reactors, such as light water reactors (LWRs).

Daigorō Yasukawa, one time president of JAPCO, reflects the continuities between Japan's wartime past and its postwar industrialization. Purged during the Allied Occupation for his involvement in providing electric power for Japan's war effort, he went on to head Japan Atomic Energy Research Institute (JAERI) at Tōkai-mura and later JAPCO. As Yasukawa so succinctly put it,

> We import the finished product not the principle. . . . At this early stage in the import of reactors we are not much concerned with Japanese scientists who do basic research. . . . Our job can be done best through private companies with as little government interference as possible.[68]

In the United States, the AEC contracted with universities for research projects, as well as building up existing university-operated national laboratories such as the Argonne National Laboratory and Lawrence Radiation Laboratory.[69] In Japan, contracts tended not to be placed with universities. Instead, public corporations were created to foster research in areas not yet ripe for commercial development but which, nevertheless, required "managerial efficiency and productivity," and closely followed the government line. By the late 1960s, the number of public corporations peaked at 113.[70]

Public corporations consist of a corporation chartered to undertake a specific public purpose with public property. They are established in order to perform the business of the state efficiently by combining the public sector with the market economy. These corporations are under the control of ministers through appointed directors and financial supervision. For this reason, they are similar to government agencies.[71]

In May 1956, the Science and Technology Agency (STA) was launched as an external bureau of the prime minister and given the task of managing Japanese nuclear R&D and of providing staff and other support for the commission. The JAERI, a public corporation, was created that same year. The Institute, located at Tōkai-mura, about 70 miles north of Tokyo and near the city of Mito in Ibaraki prefecture, was to conduct R&D. The area was, coincidentally, the homeground of Commission head Shōriki.[72] The officebearers of the Institute consisted of Daigorō Yasukawa as president, a vice-president, five directors including Sagane, and two superintendents.[73] Sagane worked at JAERI from June 1956 to September 1959, a total of three years and three months. The first half of his time was spent as a director in charge of planning, and the latter half was as deputy head director and head of the Tōkai Research Institute within JAERI.[74]

JAERI was established in the hope that a Japanese research reactor could be designed and built. It received approximately 60 percent of the government's atomic energy budget in the early 1960s, making it the major component of the atomic energy program. By 1967, staff would number 2000. The Institute aimed to develop self-sufficiency in atomic energy, whereas industry advocated importing technology through licensing.[75]

Sagane was an insider involved with the two principal groups behind the development of power reactors: (1) the nuclear power industry and electric power companies (MITI–industrial complex); and (2) the STA and public corporations for research such as JAERI. This pluralistic structure began in 1957 with the decision to introduce the British Calder Hall reactor. This marked the beginning of a split between groups arguing for domestic R&D, and those urging for the introduction of foreign technology. One could consider the establishment of STA in 1956 as one of the origins of this pluralistic structure. STA's research-oriented approach to nuclear power development, principally through JAERI, precluded the option of introducing a commercial power plant reactor.

In March 1956, Sagane was made a JAEC consultant and on June 26, 1956, he became a JAERI director. In that capacity, he had dealings with government, power utilities, and large American firms. With his technical knowledge and experience of dealing with foreigners, he was one of the few technocrats able to negotiate contracts with companies. On December 4, 1956, it was announced that JAERI had awarded a contract to the American Machine and Foundry Company subsidiary AMF/Atomics, Inc. to build a 10-MWe heavy-water-type reactor. It was hoped that the reactor would be used mainly for research and training of nuclear engineers and technicians. A photograph showing Sagane with Ichirō Ishikawa (member of the JAEC) and AMF president Morehead Patterson pointing to an artist's impression of the reactor, appeared in *The New York Times* the day after. The reactor was the fourth such reactor that AMF would build outside of the United States, the others being in Hamilton (Ontario), Amsterdam, and Munich.[76] Reactor technology was fast becoming an export industry.

Sagane's most important achievement at JAERI was the introduction of an experimental light-water power reactor, the Japan Power Demonstration Reactor (JPDR), a 12.5-MWe boiling water reactor made by GE, United States. This was in contrast with the choice of a British Calder Hall reactor, which was introduced by Japan's power utilities. LWRs were seen as heralding the new age of the power reactor. In September 1956, Sagane was made an STA atomic energy investigator and from October 15 to December 19 visited France, Germany, United States, and Canada in that capacity.[77]

The Japan Atomic Industrial Forum (JAIF) was formed in 1956 and the Japan Atomic Fuel Corporation (JAFC) in 1957. Sagane was not only involved with atomic energy policy but also participated in its actual implementation. A small American-designed and built JRR-1 research reactor (50 kW) was completed in 1957 at JAERI—the very first in Japan.[78] The year 1957 was an especially busy one. In March, Sagane was made a member of the Reactor Earthquake Countermeasure Sub-committee of JAEC. A few months later he was appointed director of the Radioactive Isotope Association and from July 1, the section chief of nuclear engineering at the Tōkai Research Institute, head of the preparatory section for the power reactor, and head of the planning and investigatory section.[79]

At a Cabinet meeting on November 15, 1957, it was decided to establish a Science and Technology Cabinet Members Forum in which members representing 11 ministries including STA, Ministry of Education, MITI, and the Economic Planning Agency could discuss matters concerning science and technology. At the forum, it was proposed to promote the expansion of science and engineering in universities and to have what ultimately became, in February 1959, the

Science and Technology Council—an advisory body to the prime minister.[80] The prime minister was, by law, required to heed the advice of the Science and Technology Council unlike that of the JSC. As a result, the Science and Technology Council had power that matched or rivaled that of Cabinet, a situation unprecedented since the war.[81] This was not surprising given that the chairman of the Science and Technology Council was the prime minister, and council members consisted of the ministers of Finance and Education, the directors of the Economic Planning Agency and STA, the president of the JSC, and three technology specialists appointed by the prime minister.[82]

Thus, despite the large number of scientists willing to advise the government, what emerged was a definite hierarchy of scientific advisory bodies, which reflected the influence of the power triumvirate of government, big business, and the bureaucracy. The Science and Technology Council was created as the highest government advisory body on science and technology policy, which would act as a type of buffer between the government and sometimes hostile JSC. Sagane was, therefore, in a particularly influential position, as many of his colleagues on the Science Council had been effectively isolated (figure 6.1).

Figure 6.1 Hideki Yukawa, Ryōkichi Sagane, and John Cockcroft in Hakone National Park, Japan, November 22, 1958.

In September 1957, Sagane became a member of the JSC Special Committee for Atomic Energy, helping to bridge the gap between JSC and the government. The following month, he was made a consultant to the Ministry of Foreign Affairs. When it came to atomic energy, international issues could not be avoided. Less than two weeks later, he was made deputy head of the board of directors of JAERI. In October, he also became head of the Tōkai Research Institute. The day after Christmas, he was made a consultant to the data examination committee of the delegation to England of JAPCO, and in March 1958, appointed part-time consultant to JAPCO itself.

Sagane's work combined overseas travel, administration, and providing technical advice. From January 25 to April 11, 1958, Sagane once more traveled overseas to England and the United States.[83] In July, he took on a number of committee memberships of the JAIF: consultant, planning committee, Cabinet legislation committee, and its atomic energy committee. In August, he became a specialist member of the JAEC.[84] In December 1958, he headed the Tōkai Research Institute, with responsibility for overseeing the construction of the JPDR. On April 11, 1959, Sagane's time was further stretched when he became a member of the Shipbuilding Technology Council. This was a sign of things to come. The Japan Nuclear Ship Development Agency (JNSDA) was established in 1963 to construct "Mutsu," Japan's first nuclear-powered commercial vessel.[85]

It was clear that Sagane was spreading himself too thinly. In September 1959, he resigned as deputy head of directors of JAERI and in December, became a full-time consultant to JAPCO, where he was involved with the construction of Japan's first, full-fledged, LWR power reactor. His gift of understanding technical matters and even U.S. culture, helped him in the negotiations to introduce American technology. As the JPDR had been a first for Japan, Sagane's experience was invaluable. His role in transferring the cyclotron from Berkeley to Tokyo in the 1930s seemed to be repeating itself on a much larger scale.

Despite protests from the scientist–critics, it was decided that JAPCO would purchase two commercial nuclear power plants of around 200 MWe, one from the United Kingdom and one from the United States. The JAEC recommended the importation of a British magnox-type, improved-Calder Hall reactor, made by General Electric, United Kingdom. JAPCO was formed to act as the semi-governmental owner–operator of the reactor. At the time, it seemed to be the only proven reactor, the world's first large-scale nuclear power reactor. In its original form, it had the dual purpose of producing both plutonium and electricity to satisfy weapon and energy needs. In December 1959, JAPCO obtained government approval to construct a modified version in the form of a 166-MWe magnox reactor, located at Tōkai-mura, Ibaraki-prefecture. Due to concerns about earthquake-proofing the Calder Hall reactor, construction took six years, ending in July 1966, at an approximate cost of US$120 million. It commenced supplying electric power to Tokyo Electric Power Company in late 1966, two years behind schedule. It was Japan's first commercial-scale reactor.

Shortly after the decision to opt for the gas-cooled magnox reactor, the United States announced a new policy of guaranteeing a supply of enriched uranium for US LWRs. The United States pursued LWRs, which use enriched uranium and were developed by General Electric (boiling water reactors, BWRs) and Westinghouse (pressurized water reactors, PWRs). Japanese experts have generally believed that LWRs were a better choice than other reactors on offer at the time. As a result, all of Japan's commercial reactors have been LWRs, especially BWRs as opposed to PWRs. This perception was helped along by good relations between General Electric and Tokyo Electric Power Company, and between Westinghouse and Kansai Electric Power Company, which dated back to the Meiji period.

With the installation of the Calder Hall reactor at Tōkai-mura and the establishment of the American JPDR, the nuclear power industry was getting a firm footing. However, given developmental problems that were being experienced overseas, initial excitement gave way to a more sober, realistic view of difficulties, which would have to be faced. Changes in attitude and the world slowdown in the nuclear industry coincided with a lowering of the cost of thermal power generation, dating from around the time of the Second Geneva Conference in September 1958[86]—almost as soon as the safety checks of the plans for the JAPCO Tōkai-mura power plant had been completed and the go-ahead certain to be granted; there was an announcement made by the then head of the AEC, Nakasone, that the long-term atomic energy generation plan would have to be revised downward in view of the changes in the enthusiasm for nuclear power.[87]

In May 1960, Sagane became part-time JAPCO executive director and in May 1961, full-time executive director. Around 1961, the JAEC began to put together a long-term plan for the development of atomic energy. Sagane was in favor of research into uranium enrichment as a national project, whereas others felt that technology for the use of plutonium in reactors should be promoted.[88] He was soon buoyed by the successful operation in October 1963 of the JPDR at

JAERI. By November 1965, commercial nuclear-powered electricity was generated for the first time in Japan. The JPDR provided training for researchers tackling problems at commercial nuclear power plants. It was used for tests and experiments with LWRs, testing of fuel materials, and research for application of LWRs to ships.[89]

In 1965, JAPCO decided to import a commercial-scale 322-MWe American GE BWR as its No. 2 Reactor, at Tsuruga, Fukui-prefecture.[90] It was completed in March 1970, at a cost of US$90 million.[91] In May 1966, Sagane became vice-president of JAPCO and became even more involved in the operation of the improved Calder Hall power plant, as well as the introduction of the BWR.[92] The experience gained at JAERI enabled Sagane to competently oversee construction of JAPCO's Tsuruga power station. He coordinated all the detailed plans for the plant building, which housed the GE 357-MWe BWR from the United States. The BWR power plant was completed in 46 months, in record short time for such construction.[93]

Meanwhile, the private sector, too, had been busy. By March 1971, Tokyo Electric had started a station using GEBWR technology in Fukushima. Kansai Electric had an operation in Mihama going four months earlier using Westinghouse PWR technology. However, the rush to commercialize nuclear power (especially LWRs) meant that a strong R&D base was not first established, and an over-reliance on American know-how was the result. This dependence on foreign technology had frustrated JAERI research workers from as far back as 1963, when an industrial dispute protesting the lack of research direction occurred. The workers complained that the government seemed uninterested in basic research. Reactor development was heavily funded, whereas staff training and research received relatively little. Scientists called for more cooperation between their activities and reactor development. As a result, there was a change of top management and more authority devolved to heads of department.[94] A policy was developed over the years for the government to develop new types of reactors at JAERI, and for the private sector to take responsibility for proven reactors.

Some of Japan's most powerful corporations came to have a stake in the continued existence of the nuclear industry.[95] These included: (1) Mitsubishi Atomic Power Industries; (2) Tokyo Atomic Industrial Consortium (Hitachi-Fuyokai Group); (3) Nippon Atomic Industry Group Company (Mitsui, including Toshiba); (4) Sumitomo Atomic Energy Industries; and (5) Daiichi Atomic Power Industry Group (Furukawa and Kawasaki).[96] The first three of this group of five have been major contractors of nuclear facilities. The JAFC became the

Power Reactor and Nuclear Fuel Development Corporation (PNC) in 1967. The Ministry of International Trade and Industry (MITI) took on the responsibility for actual nuclear power generation.

V NUCLEAR FUSION

Despite Sagane's commitment to the government line, he was aware of the need for basic research. As a physicist-cum-bureaucrat, there was a tension between his desire for truly independent action, and his being a servant of the public. This is clear in his attitude to nuclear fusion, another way of harnessing the energy of the atom. Research into fusion in Japan began around 1955, and in April 1958, an expert committee was formed with Yukawa as chairman. The committee served as advisers to the JAEC. Sagane, along with Seishi Kikuchi (a member of the JAEC), participated enthusiastically. Kikuchi was concerned that Japan might be left behind. Chairman Yukawa presided over debate between two strategies for fusion research: an A-plan, of pursuing fundamental research in plasma physics, and cultivating and training future personnel; and a B-plan, which involved making a high-temperature plasma machine, similar in scale to the C-stellerator at Princeton, and developing technology along with research. The A-plan would principally involve the Ministry of Education whereas the B-plan would mainly be centered on the STA. Sagane was a committee member with divided loyalties, being as he was both a deputy head director of JAERI (an institution where the B-plan would take place), and a former university-based academic and researcher who had been involved with large-scale experiments.[97]

In August 1959, Sagane also took on membership of the JSC Special Committee for Nuclear Fusion. The Science Council had purposely created their own committee in order to widen the debate beyond the JAEC. Kōji Fushimi was chosen as chairman and other members included Yukawa and Sakata. The JSC Special Committee, not surprisingly, recommended the A-plan under the Ministry of Education, with a plasma research institute as its central facility. The Committee hoped to delay any implementation of the B-plan. The JAEC Nuclear Fusion Expert Committee was split over the B-plan. The decision was left in the hands of Kikuchi, Yukawa, Sagane, and Fushimi. Kikuchi was keen on the B-plan whereas Sagane preferred the A-plan. Sagane felt that the B-plan was premature, especially as the requisite researchers did not exist. A report at the end of 1960 eventually led to the establishment of the Institute for Plasma Physics in 1961, with Fushimi as its first director. Sagane was not chosen director as he had been elected chairman of the Special Committee for Nuclear Fusion in 1960, and it was thought that the chair should not be director also.[98]

Sagane's involvement with fusion, nevertheless, continued with his participation in the Institute's management committee from July 1961. At the September 1961 meeting, a document authored by Sagane and Fushimi was tabled, which focused on collaborative research. Unlike the Institute for Nuclear Study at Tokyo University, where a specific machine was built and was open for use by researchers throughout Japan, in the case of the Institute for Plasma Physics, equipment for producing a high-temperature plasma would be designed, constructed, and tested by an inter-university group. Sagane's wife was from a wealthy "zaibatsu" family from Nagoya, the Okayas. Family connections facilitated the construction of equipment and facilities.[99] This was but one example of how women and their networks were important to the public lives of physicists. Such traces are often difficult to find in biographical accounts, but it is likely that the considerable wealth and connections of wives of physicists such as Hideki Yukawa, Shōichi Sakata, and Sagane, benefited their respective spouses' careers.

Sagane made many other efforts to promote plasma physics and fusion research until his death in 1969. The B-plan would eventually be acted upon when the STA commenced research into fusion in 1968.[100]

The prime minister's office, Economic Planning Agency, Ministry of Finance, and MITI attract many such researchers to their ranks, especially to work in specialized technical departments. These people are sometimes disparagingly called "goyōgakusha" ("government scholar"), meaning that they prostitute themselves by being employed directly by government. To other physicists, businessmen and bureaucrats may have appeared "effete," but this was a rather crude recognition that different "policy cultures" existed with different political and social interests, different institutional bases and traditions: the academic, bureaucratic, entrepreneurial, and civic policy cultures.[101] Most physicists were steeped in academic culture and found the values of other cultures foreign and incompatible. Although they engaged in civic culture as public men, it was in academic culture that they felt most comfortable. But was it the best place to implement the policies they helped formulate? In 1971, there were about 25,000 research scientists employed at 90 national government research institutes. It appears that the more technocratic a field is, the greater the potential influence of the intellectual employed.[102] Sagane is but one example of physicists who could make a difference. His career reflects the ascendancy of the businessman, bureaucrat, and entrepreneur, and a new vision of the public man as one immersed in the marketplace.

CONCLUSION

The Japanese Prime Minister Ikeda once remarked that government is like the captain of a ship and "zaikai" ("the business world") provided the compass.[103] The beginnings of the Japanese civilian atomic energy program reflect strong ties between the government and business, and between the United States and Japan. This strong nuclear alliance between the United States and Japan has continued to the present-day. Indeed, the path of development of nuclear power followed by Japan can be said to have been a result of the dynamics of internal cooperation and conflict. Out of this conflict reemerged a pluralistic structure—the MITI–industrial complex versus STA and public corporations—which is reminiscent of the wartime, factionalized organization of Japanese science and technology.

The history of the development of nuclear power in Japan also reflects the close collaboration between business, government, and technocratic scientists. Ryōkichi Sagane was one player in this collaboration. He was particularly well qualified for the international nature of his work after having spent lengthy periods of study abroad. Sagane combined the roles of researcher, technocrat and policy-maker, administrator, and project manager.

Sagane's career suggests that the process of technology transfer has been neglected as a subject of historical study.[104] It, too, required an expertise on a par with that of Yukawa and Tomonaga. The ability to acquire and absorb information, and then take appropriate action to utilize it, were much-needed technology transfer skills that experimental physicists had mastered. From the viewpoint of the United States, transfer of nuclear technology enabled longstanding relationships between companies to continue, and for new markets to be created. It enhanced America's image as a benevolent nation, and nurtured the U.S.–Japan alliance. For Japan, it provided a shortcut to nuclear power, improved technical capabilities in industry, and the concomitant advances in technical training and education. Sagane was but one actor in this process.

The establishment of the Science and Technology Council in 1959 as a high-level decision-making body was a result of the enthusiasm for atomic energy. The government was all too aware of the need for a policy-making body, which would have its interests foremost on its agenda. It signaled the beginning of the decline of the JSC. Despite the pressure to "toe the government line" and introduce foreign technology for the commercialization of nuclear power, the long-term prospects of fusion and the lack of experts convinced Sagane that a basic research-oriented infrastructure should be established first. This may have appeared "two-faced" to physicists who had always advocated such an approach, but to Sagane it was an extremely rational decision.

Sagane was a far less visible public man than his more illustrious physicist–colleagues and well-known father, but his role in helping to establish this research base has been very important. This was as part of "public" corporations such as JAERI and "public" enterprises like JAPCO. Sagane's experience reflects how the "public" came to be associated with public authority rather than public discussion as embodied by Sakata and Taketani. His activities were part of much larger R&D activity, for which in FY1965, the nation's total expenditure amounted to 430 billion yen (US$1.2 billion). This figure rose to 1200 billion yen (US$3.3 billion) in FY1970, the same year that JAPCO's first large-scale LWR started operation at Tsuruga and one year after Sagane's death.

The following chapter looks at a member of the next generation of physicists, Satio Hayakawa, who helped to foster Japan's research capability in space and fusion research and accelerator physics. He continued the public role that his teacher Tomonaga, and Nishina, the teacher before him, had carved out. His activities also show how the Ministries of Education and Finance have had the final say in the funding of major projects. What is emerging is a picture of a policy-making structure consisting of the LDP, bureaucracy, and business, with academia forming a major interest group. The divisions between these different sectors is often blurred. Physicists, like Sagane, span both academia and the bureaucracy. The business leader Matsutarō Shōriki is an example of how the private sector has gained access to the bureaucracy and government. The experiences of Sagane, and, in the next chapter, Hayakawa, testify to the need for close interaction with politicians and bureaucrats in big science. The story continues.

CHAPTER 7

SCIENCE ON THE INTERNATIONAL
STAGE: HAYAKAWA

In September 1945, the young physicist Satio Hayakawa joined a
group headed by Osamu Minakawa (Central Meteorological Bureau)
and Ryōkichi Sagane (Tokyo University) to inspect the impact of
the atomic bomb at Nagasaki. Hayakawa's job was to measure radioactivity.[1] The American physicist Richard P. Feynman whom he would
later befriend, had worked on the Manhattan Project that ultimately
led to the destruction at Hiroshima and Nagasaki. Ironically, Feynman
and other American physicists sought to help Japanese physicists
rejoin the international community of science.

Why focus on Hayakawa? A weekly magazine, the *Asahi jaanaru*,
ran a profile of Satio Hayakawa in March 1961, written by one of his
physicist colleagues. The article described the pipe-smoking, athletic,
and fairly tall young professor as "Mouth, hands, legs all skillful."
Written by Nobuyuki Fukuda, it related how his friend of some
15 years

> habitually speaks of physics, the research system and organization.
> I haven't heard of him speak of anything else. . . . Hayakawa is both an
> excellent scholar and wonderful organizer. Among Japanese this sort of
> person is rare.[2]

Hayakawa's style of physics reflected his personality and character. He
was quick to assess situations, whether it be on the research front or in
the committee room. His interest in the "new" was indicative of his
impatience with the old. He was strong in theory and quick to understand and embrace new concepts. His wide knowledge of physics and

interest in many fields facilitated this. But he was not a reckless scientist who only followed fashions. He was highly practical and placed great importance on experiments and observations.[3]

In previous chapters, we examined the activities of physicists known affectionately as the "daibosu" ("big bosses") of the Elementary Particle Theory Group—Hideki Yukawa, Sin-itirō Tomonaga, Shōichi Sakata, and Mituo Taketani. They were followed by a group of physicists called "chūbosu" ("middle bosses"). The middle bosses graduated from university around the time of World War II. This group includes physicists such as Seitarō Nakamura, Takeshi Inoue, and Nobuyuki Fukuda, as well as a sub-group of physicists who completed university studies at the end of the war or soon after: Satio Hayakawa, Yōichi Fujimoto, Hiroomi Umezawa, and Eiji Yamada.[4]

This final chapter focuses on Hayakawa's activities, for they provide a useful window from which to view the continuities and changes occurring in the roles played by physicists in policy-making and in the broader society, right up until 1970—the end of the period under study. Hayakawa's career gives us a sweeping view of the rise and decline of Japanese scientists in policy-making. He was the most outstanding of the middle bosses. In the words of a fellow "middle boss" Nobuyuki Fukuda, he was a veritable "superman."[5] Given the appalling research environment in which physicists found themselves immediately after World War II, the Japanese would need superhuman qualities to maintain their research activities. It was inevitable that they would turn to the United States. American physics beckoned. Going overseas also served to emotionally distance themselves from a defeated Japan. Postwar Japanese physicists were faced with a dilemma similar to that faced by their Meiji period predecessors. During both periods, the Japanese had no option but to internationalize, quickly. Hayakawa wrote that Japan's 300 years of semi-seclusion, geography, and language all worked against international collaboration. He was convinced of the need to overcome these barriers and promote science that went beyond Japan's own boundaries.[6]

Much big science in the postwar period required some form of international collaboration. This provided a further impetus to the already close U.S.–Japan ties, and ensured that the Americanization of physics continued on after the Occupation. Many of Japan's most outstanding young physicists crossed the Pacific to study in the United States in the late 1940s and early 1950s. The United States replaced Germany as the favored destination of physicists. Hayakawa was one of those who went. His career is briefly described in the first section of this chapter, later sections providing details of his activities in nuclear fusion, space research, and accelerator physics. The chapter also provides evidence of Hayakawa's remarkable breadth of research, his internationalism, and his talent for institute building. It is through his policy-related activities that we also see the reemergence of a multilayered policy system, referred to in previous chapters. The Japan Atomic Energy Commission (JAEC), Science and Technology Agency, Science and Technology Council, and other bodies supplanted the Science Council in areas of policy-making that physicists considered their own. Their professional autonomy was constrained but internationalism gave them a voice that could travel beyond national borders.

I A BRILLIANT CAREER

How did one become a public man, a "boss" of Japanese physics? Hayakawa was born in Niihama city, Ehime prefecture, in 1923, less than a year after Einstein's visit to Japan and amidst the boom in physics that it generated. He went through the rites of becoming a physicist by studying at the Musashi Higher School, an elite private boys school in Tokyo. He not only excelled academically but was also a member of the school "kendō" ("Japanese fencing") club.

Although most young, able-bodied Japanese men were mobilized for the war, Hayakawa graduated from high school in September 1942, and entered Tokyo Imperial University the following month, majoring in physics. He was part of the great wartime increase in science and engineering graduates in Japan. As compared to a decade earlier, the number of such graduates tripled during the Pacific War. Some were absorbed in the expansion of domestic research facilities. Hayakawa's friend Yōichirō Nambu was assigned to radar research by the Army.[7]

Hayakawa completed his studies at the end of the war in September 1945. The casualties of war deprived Japan of some of its best young talent. Hayakawa was poised to fill the vacuum that would, in time, be created by the withdrawal from the public sphere by physicists such as Nagaoka and Nishina, who would pass away.[8] The premature transfer of power to the next generation meant that young scholars such as Hayakawa were old before their time.

Even at the end of the war, Hayakawa's research interests were very wide and encompassed physics and astronomy. This was partly a result of the nature of physics at the time, his own enthusiasm for learning, and postwar pragmatism, which encouraged young Japanese to excel in a number of areas that could be potentially useful in their careers. He worked on nuclear physics and elementary particle theory, concentrating especially on cosmic rays and astrophysics. As a member of one of Tomonaga's study groups, he helped to develop the super-many-time theory and renormalization theory immediately after the war. Hayakawa's earliest papers were coauthored with Tomonaga.[9] Each week, there were meetings where work-in-progress reports were presented and relevant scientific literature discussed.

Despite the devastation in Tokyo and poor living conditions, these young physicists had plenty of intellectual stimulation with meetings of one type or another almost everyday. Immediately after the war, Hayakawa joined the cosmic-ray-experiment section of the Central Meteorological Bureau, headed by Osamu Minakawa, as an unofficial member. In late September, he was asked to join Minakawa and Ryōkichi Sagane's Tokyo University group in an inspection of Nagasaki.[10]

Hayakawa's weekly schedule consisted of spending Mondays at the Minakawa laboratory at the Meteorological Institute; Tuesdays at a Tokyo University study meeting to discuss new scientific literature; Thursdays at a Nishina laboratory colloquium and Fridays with the Tomonaga group at the University of Science and Literature (Bunrika University). Although often hungry for something to eat, his busy schedule ensured that he would at least be intellectually nourished in Tokyo and get to meet the big bosses of Japanese physics at the same time.

At the Tokyo University Tuesday meetings, Hayakawa was quick to tell the hot news in the scientific world and to ask questions of others. His major source of new scientific information was SCAP's (Supreme Commander for The Allied Powers) CIE (Civil Information and Education Section) Library in Hibiya, Tokyo, where he voraciously read newly arrived issues of foreign journals. To ensure that he was the first to read them, he would arrive at the library before it opened.[11] Even though he was cut-off from the rest of the world, through the library and its journals, Hayakawa accessed an international network.

On Fridays, Hayakawa would often sit in the front row of the Tomonaga seminar, asking the most questions and becoming one of the core groups revolving around Tomonaga. To those who attended, it appeared that Hayakawa and the other bright Tokyo University scholars monopolized proceedings. Such seminars effectively served to establish a hierarchy of intellect, to create difference and confirm who the leaders of the physics community would be. Hayakawa and his colleagues formed an elite of sorts, one which was marked by a hunger for scientific stimulation, and which aggressively asserted its intellectual superiority.

Hayakawa seemed to have an air of maturity about him, well beyond his actual years. His nickname amongst his colleagues was "old" Hayakawa and the pipe that he continued to smoke until his death was part of his international style. This maturity made it easier for him to join the physics hierarchy as a middle boss. Along with it was an expectation that he would be the next Yukawa, only better looking! His scientific work was strong in theory and reflected a very sharp mind.[12]

Hayakawa also obtained a small income by doing temporary office work at the Meteorological Bureau in October 1945, and technical work in the research institute there at the end of 1946.[13] In late 1946, Tomonaga suggested to Hayakawa that he take up cosmic-ray research, something which Hayakawa's boss Prof. Minakawa supported.[14] In January 1947, he was appointed technical officer by the Ministry of Transportation and made a member of the Meteorological Research Institute, working in Minakawa's electromagnetism laboratory. From mid-1949, Hayakawa became a lecturer in the Faculty of Science and Engineering at Osaka City University.[15]

Hayakawa was "enticed" to Osaka by Yuzuru Watase, a physicist who was intent on bringing together a group of young, outstanding physicists. The group included Yoshio Yamaguchi and Yōichirō Nambu, and after Hayakawa's departure for the United States, Kazuhiko Nishijima and Tadao Nakano.[16] Just prior to the commencement of the 1950 academic year, Hayakawa was promoted to associate professor.

In May 1950, Hayakawa left on his first trip overseas to participate in the Foreign Student Summer Project at MIT. He spent some seven months at MIT, working with Bruno Rossi at the Laboratory for Nuclear Science and Engineering.[17] Hayakawa's overseas study was part of what the U.S. State Department called the exploitation of "opportunities for the achievement of favorable international relations." As part of its general recommendations on "Science and Foreign Relations" issued in May 1950, the Department suggested that government exchange programs be encouraged, along with "the conduct of privately-sponsored programs involving international science." Not only this, but policy-makers were advised to "acquire an awareness of the scientific implications of their decisions."[18]

Hayakawa was one of the few fortunate young physicists who had the opportunity to travel overseas before the end of the Allied Occupation. In 1948, Yukawa obtained GHQ permission to accept an invitation to visit the Institute for Advanced Study at Princeton for a year during 1948–1949. Nishina visited Copenhagen in September 1949 to attend the general assembly of the International Council of Scientific Unions, and Tomonaga was invited to Princeton for about a year during the period 1949–1950. Seishi Kikuchi would also follow suit and visit Cornell University (1950–1951). The young and promising Hayakawa visited MIT where he studied nuclear theory with physicists such as Rossi (June–August 1950), later moving on to Cornell University (September 1950–July 1951). In the 1930s and 1940s, cosmic rays were studied in order to gain insight on high-energy interactions and elementary particles. Hayakawa's interest in cosmic rays thus tended to be on their ramifications for high-energy physics. But his work with Rossi led him to a new aspect—the question of the origin of cosmic rays.[19]

Hayakawa went on to England and in August 1951, France and then India, leaving India in December and returning to Japan in January 1952.[20] In his absence, he received his D.Sc. from Tokyo University in November 1951 for research on cosmic-ray high-energy phenomena. Tatsuoki Miyazima would go to Birmingham (1952–1953), Hiroomi Umezawa to Manchester (1953–1955), and Yōichirō Nambu and Tōichirō Kinoshita to Princeton in 1952.

Hayakawa returned in 1952 and that year married Michiko, the daughter of a businessman who worked for Mitsubishi Corporation. Michiko and her parents went to live in Germany soon after she was born, and only returned to Japan when she was 11 years old. She was as fluent in German as she was in Japanese. In marriage and in research, Satio Hayakawa was "international." Sometimes described by Japanese as being a "yankee boy," he was outspoken and not fearful of being different.

Following Nambu's departure for the United States, Hayakawa filled his position on the Yukawa Commemorative Hall Committee. This was Hayakawa's first opportunity to become involved in the concept of joint-use research institutes. Yukawa Hall became established as the Research Institute of Fundamental Physics in the summer of 1953 and Hayakawa became involved in the organization of research activities.

Improved U.S.–Japan relations made for rewarding scientific exchanges, especially during the Allied Occupation when Japanese physicists endured harsh living conditions.[21] Hayakawa introduced Minoru Oda to Rossi and this resulted in Oda joining Rossi as a post-doctoral student, working on a cosmic-ray air shower project during the period 1953–1956, and collaborating in later years. In 1966,

Hayakawa was instrumental in having Oda return from MIT and join the Institute of Space and Aeronautical Science (ISAS), which had just been established. Oda formed an x-ray astronomy group there. It was international exchange in the true sense in that Japanese physicists who studied abroad in the United States did sometimes come back. Yōichirō Nambu would choose not to return.[22]

Hayakawa's experience abroad confirmed to him the benefits of science as an international and progressive force, something which would enable Japan to pick itself up from defeat, make friends, and win international prestige. From July 1952 to October 1953, he was able to show his colleagues his newly acquired internationalism when he became a member of the organizing committee for the International Conference on Theoretical Physics, which was held in Tokyo and Kyoto in September 1953, under the auspices of the Science Council of Japan. Yoshio Fujioka was secretary-general and Masao Kotani assisted as secretary. It was Japan's first worldwide international scientific conference.[23] Kotani described the conference in the following way:

> It may not be an undue exaggeration to say that this conference opened the door to international exchange in science, which had been closed since the beginning of the World War.[24]

Young Japanese physicists such as Hayakawa were eager to embrace a U.S.-inspired modernity after World War II, but American physicists attending the conference, such as Richard P. Feynman, were more interested in the traditional culture that made Japan different. On arrival in Tokyo, Feynman was taken to the Imperial Hotel, a first class, Western-style hotel designed by Frank Lloyd Wright. He expressed disappointment at what he felt was an imitation of a European hotel. "We weren't in Japan; we might as well have been in Europe or America!"[25]

Feynman's disappointment was compounded by the fact that he had studied some Japanese before arriving in Japan. John A. Wheeler, who had been Feynman's thesis advisor at Princeton University, had written to American participants encouraging them to learn some Japanese, using an Army phrasebook, prior to the conference. Feynman used his newly gained prowess in Japanese to order coffee at the Imperial Hotel, much to his physicist friend Robert E. Marshak's surprise.[26]

Despite attempts to dissuade him, Feynman arranged to move to a Japanese-style hotel where he would have to "sleep on the floor" and "sit on the floor at the table."[27] Feynman enjoyed the layout of his new hotel room, complete with "tokonoma" ("decorative alcove"), and sliding doors that led out to a garden. He changed into a light-weight, blue-and-white, cotton "yukata" ("summer kimono"), and later made use of the toilet, only to find that Yukawa was using the adjacent, traditional Japanese bath! Yukawa advised Feynman that it was poor etiquette to use the lavatory when the bath was occupied.[28] Despite their mutual embarrassment, it is rather telling that both Yukawa and Feynman preferred a traditional-style hotel, even at the cost of having to share a bathroom! Indeed, Feynman resolved that "I was going to live Japanese as much as I could."[29] As one of the organizers, Hayakawa helped to make that happen.

On their way to Kyoto, where the second part of the conference was held, Hayakawa, Feynman, and Abraham Pais stopped off at Gifu city, near the picturesque Nagaragawa River where a local attraction is comorant fishing. After having heard about Feynman's Japanese-style hotel, Pais, too, was keen to try such accommodation. One of the attractions of a Japanese-style room was the ease with which people could drop by, sit on the floor, drink Japanese tea and eat sweets or crackers, and talk.[30] Indeed, Hayakawa, Feynman, Pais, and S. Nakajima did just that in a Japanese-style room at the Nagaragawa Hotel on September 17, 1953. On arrival in Kyoto, Feynman and Pais chose to share a Japanese-style room at the Hotel Miyako.

Feynman had a great interest in Japan and its culture. He developed a fondness for the country and its people during the 1953 conference and on a second visit to Japan in the summer of 1955, which was mainly spent at Yukawa Hall (figures 7.1 and 7.2). Contact with Feynman and other American physicists such as Marshak encouraged Hayakawa and his colleagues to visit the United States.

Due to the so-called Yukawa effect, many young people had been attracted to a career in theoretical physics, only to find a lack of post-graduate scholarships and professional posts. At the conference, Marshak suggested that two students be sent to the University of Rochester annually to study under a scholarship award. As a result, Tomonaga started sending talented graduate students to study with Marshak from that very year. Susumu Ōkubo, Yasushi Takahashi, and Masatoshi Koshiba were able to establish themselves as U.S.-based physicists via this program.[31]

Cosmic-ray physics was one area in the 1950s where much low-cost research and scientific exchange could be conducted. In his own work, Hayakawa linked the evolution of stars with the elementary

Figure 7.1   Back row (standing): Jun J. Sakurai, Chūshirō Hayashi, Mituo Taketani, and Satio Hayakawa. Seated: Seitarō Nakamura (second from the left) and Hideki Yukawa (center right). February 1955 in Yukawa Hall, Kyoto University.

Courtesy, AIP Emilio Segrè Visual Archives, Yukawa Collection.

composition of cosmic rays and proposed a hypothesis regarding the origin of cosmic rays in supernova.[32] In January 1954, Hayakawa took up a professorial position at the young age of 30 at the new Research Institute for Fundamental Physics (RIFP) at Yukawa Hall, Kyoto University. In August of the following year, he became a member of the RIFP Management Committee where he would rub shoulders with many senior physicists, and pursue his activities in helping to build Japan's research infrastructure. He would also again play host to Feynman who visited Yukawa Hall that summer.

Appointments to other important committees came thick and fast. In April 1956, Hayakawa became a specialist member of the Tokyo University Institute for Nuclear Study preparatory committee, and in May 1958, the JAEC. Committee memberships continued with the Science Council of Japan's Special Committee for Nuclear Research (SCNR) in December 1958, the SCNR's Subcommittee for Future Plans for Nuclear Research in July 1959, and the Cabinet's Council for Space Development in December 1960. Hayakawa also made a point of traveling overseas during these years, to Mexico in 1955, the United States from August 1956 to May 1957, and USSR in 1959.

Figure 7.2 Sumi Yukawa, Satio Hayakawa, Richard P. Feynman, Hideki Yukawa, Kōichi Mano, and Minoru Kobayasi, during a visit to Yukawa Hall, Kyoto University, summer of 1955. (Image has been reversed.)

Courtesy, AIP Emilio Segrè Visual Archives, *Physics Today* Collection.

In March 1959, Hayakawa was appointed professor at Nagoya University, specializing in astrophysics, especially infrared astronomy. In the early 1960s, with the discovery of x-ray emission from space, Hayakawa began cosmic x-ray observations, and helped to clarify the existence of high-temperature plasmas in the vicinity of the sun.[33] He became increasingly well known amongst the international physics community, and in March 1961, traveled to Austria to attend a committee for the establishment of an International Atomic Energy Agency research institute for theoretical physics.[34]

For at least the first ten years of the Kyoto-based, English language journal *Progress of Theoretical Physics*, Hayakawa was the largest single contributor of papers, with research that included the areas of elementary particle theory, nuclear physics, solid-state theory, relativity, cosmic rays, and astrophysics. During his career, Hayakawa would author over 300 papers. According to Fukuda, he was verbally "skillful" in his articulateness and great enthusiasm for research and organization. Furthermore, his hands were skillful in the sense that he helped to build organizations such as the Institute of Plasma Physics, and his "legs and feet" were skillful or nimble in that he had traveled to most places in the northern hemisphere.[35]

In 1961, the young graduate student and budding physicist Fumitaka Satō met Hayakawa for the first time. His impression was that Hayakawa seemed more like a foreigner who spoke Japanese than an actual Japanese—so international had he become.[36] Whether it was a sign of his intelligence or a trait borrowed from Feynman is difficult to say, but Hayakawa had a tendency to ask bruising questions at research seminars that often got to the heart of a topic.[37]

Hayakawa consciously developed a persona as an international scientist with a network to match. In 1963, he gave lectures on the origin of cosmic rays at the Brandeis University Summer Institute in Theoretical Physics, which was supported by the U.S. National Science Foundation and NATO (North Atlantic Treaty Organization). Other speakers at the Institute included physicists from NASA (U.S. National Aeronautics and Space Administration), Columbia University, and Iowa State University. Hayakawa's own lecture notes, along with those of the other speakers, were published the following year and widely circulated.[38]

In 1967, he received the Chūnichi Cultural Prize for his research in x-ray astronomy, and two years later, a definitive textbook on cosmic-ray physics written by him in English was published by a major American publisher. It would be for the latter that he would be best known amongst Western physicists.[39]

Hayakawa won the Asahi Prize in 1973 for theoretical work and observations of high-energy astronomical phenomena. From April 1970 to December 1971 and for two years from 1975, Hayakawa was dean of the Faculty of Science at Nagoya University. His first term as dean was at a time of great campus unrest.[40]

Around 1978, Hayakawa, mindful of the need for recognition of Japanese contributions to international physics, was instrumental in initiating, with Professor Laurie M. Brown of Northwestern University, a Japan–U.S.A. collaboration on the history of particle theory in Japan for the period 1930–1960. The project, which would effectively last for some 14 years, generated a large number of publications[41] and raised the consciousness of Japanese physicists to the importance of documenting their role, both scientific and social, in shaping Japanese and international science.

After his retirement in March 1987, he was called back to be president of Nagoya University and was reelected to a second term in 1991. He received the Japan Academy Prize in June 1991 for his achievements, particularly for his work on the supernova origin of cosmic rays.[42] He would be unable to complete the second term of office, passing away on February 5, 1992.[43]

While Hayakawa's activities in policy-making were many and were varied, his own academic research interests tended to focus his energies on particular areas, areas that due to his more international perspective would position him in the forefront of exciting new research opportunities: nuclear fusion, space sciences, and accelerator physics. It was perhaps natural that he would play a formative role in the establishment of the Institute of Plasma Physics (Nagoya University), Institute for Space and Astronautical Science (Tokyo University), and the KEK High Energy Physics Laboratory (Kō Enerugi Butsurigaku Kenkyūjo) at Tsukuba. By doing so, Hayakawa helped bring "big science" to Japan. As we saw with Sagane, this demanded certain skills of scholarly judgment, leadership, and an ability to communicate with the bureaucracy and foreign scientists. It needed an outgoing "international scientist" with good English skills, such as Hayakawa, rather than the more reclusive (although highly venerable) type of scientist as represented by Yukawa. But unlike Sakata, Hayakawa devoted his energies to making science happen rather than using science to change society.

In the 1950s, the three laboratory programs that became "bellwethers of American scientific standing were nuclear power reactors, high-energy accelerators, and fusion."[44] Japan, too, sought to develop such programs. We saw in the previous chapter how Sagane was heavily involved in the Japanese development of nuclear power. It is not surprising that Hayakawa would become involved in accelerators, fusion, and space research.

II HAYAKAWA IN FUSION

At the beginning of the 1950s, scientists were the main force behind the fusion program in the United States and the search for fusion reactors. There was, however, no established discipline of fusion physics. There was little by way of known theory on plasmas, the so-called fuel for fusion reactors. Interested scientists came from a number of disciplines. Astrophysicists could understand the cool plasmas of interstellar space and the hot plasmas of the interior of stars. Specialists in particle accelerators knew about designing magnetic fields and the potentialities of power supplies. Cosmic-ray scientists had an understanding of the behavior of charged particles in complex magnetic fields. Developments by scientists in such fields helped stimulate interest in the scientific problems posed by fusion. Japan was no exception and Hayakawa's research conveniently spanned all three.[45]

In the early years, a fusion reactor was considered as having considerable military potential. In 1953, the U.S. Atomic Energy Commission's support of research into controlled thermonuclear fusion amounted to less than one million dollars. This increased rapidly in the following years to reach the amount of ten million dollars in 1957. During the first few years, scientists at Los Alamos, Berkeley, and Princeton were optimistic that a path to realizing a controlled thermonuclear reactor would be found. There was keen competition shown by Princeton University, Los Alamos Scientific Laboratory, Livermore Laboratory run by the University of California, and the Oak Ridge National Laboratory for the large amounts of funding being made available for U.S. fusion research. The "stellarator" at Princeton, the "pinch" at Los Alamos, and the "mirror" at Livermore all achieved some success at plasma confinement. By the end of 1957, the success of the pinch machines suggested that they could be scaled up to make fusion reactors.[46]

International interest in fusion was stimulated by comments at the first International Conference on the Peaceful Uses of Atomic Energy at Geneva in August 1955, which was attended by a 17-member delegation from Japan that included the physicist Yoshio Fujioka.[47] Homi J. Bhabha speculated that "a method will be found for liberating fusion energy in a controlled manner within the next two decades."[48] This triggered comments by the U.S. Atomic Energy Commission chairman, Lewis L. Strauss on the progress of American research into controlled thermonuclear reactions. The optimism concerning fusion was further reinforced by comments by members of the fusion group at the Atomic Energy Research Establishment at Harwell. Bhabha's comments provided the Japanese with confirmation of the possibilities opened up by fusion research. Minoru Okada of Osaka University was traveling in Europe at the time. Hearing the news in Zurich, he became interested in the possibility of pursuing research into nuclear fusion in Japan.[49]

When Okada returned to Osaka University around November 1955, his colleagues persuaded him of the importance of commencing basic fusion experiments in Japan. On June 26, 1956, a public experiment was conducted at Osaka University. Those who attended formed the basis of the Super High Temperature Research Group. Kōji Fushimi was chosen as chairman and the first regular meeting was held on October 26.[50] In late 1956 at the Faculty of

Engineering at Osaka University, the Okada Laboratory group conducted experiments on large current discharge.[51]

By 1957, technical problems had beset projects in the United States and elsewhere. While dampening the spirits of scientists, it did not dampen the enthusiasm of the Atomic Energy Commission (AEC) to use nuclear fusion as the centerpiece of the U.S. exhibit at the 1958 Geneva conference, in many ways the international fair for nuclear technology. The display included models and experiments from the abovementioned U.S. institutions. There was a growing realization, however, that a working reactor could not simply be put together, but that "basic" research in theory and experiments would need to be conducted. Up until 1958, the year of the Second International Conference on the Peaceful uses of Atomic Energy, details of the U.S. program were kept secret. After considerable lobbying by scientists, work on fusion research was declassified by the AEC on the eve of the Geneva conference. This was in the hope that greater developments might be made if there was a freer exchange of ideas amongst scientists. The United States also obtained political mileage out of this gesture of generosity to other nations.[52]

Despite the impressive U.S. fusion display, it became apparent to all the delegates of the second Geneva conference that there were immense technical difficulties to be overcome in the development of nuclear fusion. Wasteful duplication of research had occurred and international cooperation was one way of avoiding this.[53] Declassification enabled fusion research to take on more characteristics of academic research. Universities and scientific societies began to wield more influence on the direction of research. Declassification also allowed interested industrial concerns to stake a claim in the area as well. General Electric was quick to do so.[54] Public disillusionment with nuclear fission encouraged utility companies to look at fusion as an alternative energy source.

In September 1955, Hayakawa, then associate professor at the RIFP, attended the International Conference on Cosmic Rays in Mexico. He wrote back to Yukawa, then director of the Institute, of the need to commence nuclear fusion research in Japan:

> Though in Japan, people are giving their attention rather to nuclear fission energy, I think if we start the study of fusion reaction now, we would easily catch up with the developed countries. The United Kingdom and the Soviet Union and particularly the United States seem to be conducting their studies secretly. If we do this study openly, we will be able to compare favorably with these countries. What is important is how we should organize the cooperation of people in various fields.[55]

"Catching up" with the West, open access to research, international prestige, and the need to organize research for effective cooperation were themes that continued to provide a rationale for Hayakawa's policy-making activities. With such hopes in mind, Hayakawa encouraged nuclear physicists and astrophysicists to meet at RIFP in February and October of 1955. They discussed whether the nuclear reactions in stars could be reproduced in a controlled manner on Earth. One of the first open discussions on the use of thermonuclear reactions in Japan was in April 1956, at a week-long meeting of the "Chōkōon kenkyūkai" ("Super-High Temperature Research Meeting"). The meeting was conducted at the RIFP and organized by Mituo Taketani, Seitarō Nakamura, and others. Discussion topics included the question of what properties of matter at the super-high temperature would be required for thermonuclear reactions.[56] That same month, Ivor V. Kurchatov presented an important lecture at the Harwell Laboratory in Britain on the progress of Soviet research into pinched discharge.[57] Soon after, Japanese scientists conducted experiments on condensed discharge, in an attempt to obtain nuclear fusion.

From this time on, a number of projects centering around the construction of machines to produce high-temperature plasma were proposed in Japan. The cost of such an apparatus would have been prohibitive for any one particular institution. On October 11, 1956, at a regular meeting of the JAEC, it was decided to establish a committee for the peaceful uses of nuclear fusion. The Committee on Nuclear Fusion was chaired by Hideki Yukawa. Seishi Kikuchi, a member of the Commission, did much of the organizing and the secretariat was run by the Atomic Energy Bureau. The Committee hoped to build a machine comparable to the American machine Stellarator B. In order to put together a proposal for an appropriate machine, a special study group was formed, headed by K. Yamamoto (Nagoya University).

There was considerable interest shown in nuclear fusion by Japanese scientists. A forum on the topic was organized on February 6, 1957 under the auspices of the JAEC. Participants included Ryōkichi Sagane, Minoru Okada, Kōji Fushimi, Seitarō Nakamura, Mituo Taketani, and Chūshirō Hayashi. It was decided that a society would be created, specifically for the new field of fusion and plasma physics, for which Yukawa would be one of the main organizers. By autumn, a national body of researchers, called the Nuclear Fusion Forum ("Kakuyūgō Kondankai"), had been established with some help from the Atomic Energy Bureau of the Science and Technology Agency.[58]

There was widespread interest. At the first meeting of the Forum, those invited to attend included the following: JAEC representatives Ichirō Ishikawa, Yoshio Fujioka, Hiromi Arisawa, and Hideki Yukawa; Japan Atomic Energy Research Institute (JAERI) directors Ryōkichi Sagane and Asao Sugimoto; Minoru Okada and Kōji Fushimi (both from Osaka University); Seitarō Nakamura (Tokyo University); and Chūshirō Hayashi (Kyoto University). Mituo Taketani (Rikkyō University) was absent. The meeting was chaired by Yukawa. By the second meeting, the number of participants had swollen to include a further 23 persons. Due to the interest shown, it was decided to appoint a number of regional secretaries: Seishi Kikuchi (Institute for Nuclear Study) for the Kantō area; Ryōkichi Sagane (JAERI), Kōji Fushimi (Osaka University), and Satio Hayakawa (RIFP) would be responsible for the Kansai area.[59]

Despite this interest, the budget for fusion research in 1957 in Japan was only 900,000 yen (US$2500)[60] (which was provided by JAERI) and an Osaka University scientific grant-in-aid of 5,000,000 yen (US$13,889). It was decided that a request for increase in funds for 1958 would be made, and an expanded R&D program would be pursued in 1959, in readiness for large-scale research that would commence in 1960.[61]

The management of the Nuclear Fusion Forum differed from what occurred in the case of fission research. Although the Forum operated under the guidance of the Atomic Energy Bureau of the Science and Technology Agency, deliberations were made known to researchers and close links maintained with the Ministry of Education and the Science Council (see figure 7.3).

After the second meeting of the Forum, the government set up a special committee to formulate policy for fusion research. Meanwhile, researchers made preparations to establish an independent Fusion Forum. At the end of July 1957, a Super High Temperature Research Group symposium was held at Osaka University. This was followed by the launching of the new Fusion Forum at a symposium at the Japan Academy building in Tokyo on February 10–11, 1958.[62]

The results of a pinch experiment at Osaka University were reported at a nuclear power symposium in February 1958. This attracted much interest as it came shortly after the reported "success" of the ZETA (Zero Energy Thermonuclear Assembly) pinch experiments in the United Kingdom in late January. The Osaka experiments served to heighten Japanese interest in nuclear fusion. Gorō Miyamoto (Tokyo University), Seishi Kikuchi (INS director), Gotō (Electrical Research Institute director), and Okada (Osaka University)

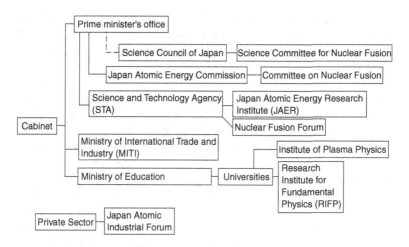

Figure 7.3 Organization of fusion research, ca. 1950s–1960s.

Adapted from Akira Oikawa, "History and Organization of Fusion Research and Development in Japan," *Fusion Nuclear Technology*, vol. 17, no. 2 (March 1990), pp. 232–35, esp. p. 233.

were requested to appear before the House of Representatives Special Committee for Policy for the Promotion of Science and Technology on February 29, 1958. After the hearing, the government took measures to set up a nuclear fusion committee within the JAEC, to be headed by Yukawa.[63] In addition to the activities associated with the Forum, researchers in Tokyo who were interested in fusion research participated in a group organized by Gorō Miyamoto, professor at Tokyo University. A small group was also organized in April 1958 by the Department of Physics in the Faculty of Engineering at Nihon University in Tokyo.[64]

International interest in nuclear fusion at the Second International Conference for the Peaceful Uses of Atomic Energy in 1958 added further momentum to discussions in Japan over the possibility of constructing a Japanese machine. Reports of the state of fusion research throughout the world were made at the conference in Geneva. Fusion research in Japan had, to that date, been funded to the extent of 80 million yen (US$222,222), two thousandths of the U.S. amount.[65] After several meetings, the Committee for Nuclear Fusion proposed two types of study groups to conduct feasibility studies: (1) a group to design an original Japanese machine, and (2) a group to prepare for the construction of a medium-sized machine based on overseas designs. Each strategy became known as the A-plan and B-plan, respectively. The Atomic Energy Bureau was in favor of Plan B. The majority of the members of the Fusion Committee preferred to give first priority to Plan A, but the Committee indicated that Plan B should also be looked into. An interim proposal was issued on March 30, 1959.[66]

Plan A would involve university-based research and joint-use research institutes. This was the domain of the Science Council. The Forum wanted appropriate representation in the Science Council of Japan to investigate such matters. At the April 1959 general assembly of the Science Council, it was decided to form a Special Committee for Nuclear Fusion. This came into being on May 14, 1959.[67] The Committee was chaired by Kōji Fushimi (sometimes written Kōdi Husimi) of Osaka University. The two secretaries consisted of Eiichi Kawasaki of Nihon University and Tarō Kihara of Tokyo University. Initial members of the Special Committee included Sakata, Yōichi Fukushima, Fushimi, Sagane, Nobuyuki Fukuda, Yōichi Fujimoto, Ryōgo Kubo, Yukawa, Minoru Okada, Hayakawa, Chūshirō Hayashi, and Gorō Miyamoto. It is no coincidence that the Committee consisted of many of Japan's leading physicists. There was, obviously, some overlap of committee membership. Persons such as Yukawa simultaneously held the positions of Fusion Forum chairman, Fusion Committee chairman, and member of the Special Committee for Fusion.[68]

The JAEC Fusion Committee approached JAERI to examine Plan B. In response to this request, JAERI established its own committee to conduct investigative research into a medium-sized machine. After much debate, the Fusion Committee chairman Yukawa consulted with JAEC member Kikuchi, Special Committee for Nuclear Fusion chairman Fushimi, and JAERI director Sagane. A decision was made to recommend inclusion of the B-plan as well as the A-plan, while placing emphasis on the role of universities.[69]

At the third and fourth meetings of the Special Committee in July and August 1959, there was discussion as to whether to (a) locate a fusion research section at JAERI or (b) establish a national research institute or university-attached research institute. Given the membership of the committee, it is not surprising that the latter was chosen.[70] Many scientists, somewhat upset after their experiences with the government and business over the tendency to import reactor technology rather than develop it themselves, concluded that the Atomic Energy Commission's plans to build a big machine were not feasible, and that a central research institute should instead be formed for the study of plasma physics. The Science Council committee subsequently put together such a proposal.[71] This proposal to establish an Institute of Plasma Physics was subsequently submitted to the government as a recommendation of the twenty-ninth general assembly of the Science Council.[72] A preparatory subcommittee for the establishment of the Institute of Plasma Physics was created at the August 2, 1959 meeting. The sub-committee was chaired by Tarō Kihara. Due to the reelection of the Science Council, the sub-committee continued until October 1959, after which it awaited the reformation of a new Special Committee for Nuclear Fusion. An interim sub-committee continued from November 17, 1959 to February 17, 1960, with Sagane as chairman. This was followed by a seven-person committee, which continued until February 25, 1961. This committee included Hayakawa and Sagane.[73] Although it was decided that the Institute of Plasma Physics would be attached to Nagoya University, questions regarding research content and organization were examined by the preparatory committee and study groups of the Special Committee for Nuclear Fusion.[74]

The bureaucratization of fusion research and the influence of universities on the direction of research tended to channel it toward basic science. Although the goal of obtaining thermonuclear temperatures still remains, this is somewhat secondary to the aim of coming to an understanding of the physics of fusion plasmas. Plasma theory has, however, tended to involve the application of established laws that are applied to very complex phenomena.[75] In contrast to the interdisciplinary nature of the scientists involved in fusion in the 1950s, in the 1960s, there emerged a community of plasma specialists who were graduates of programs in plasma physics established in the late 1950s. In Japan, for example, programs in fusion science were established at several universities in 1959, in line with the recommendations of the A-plan.[76] Throughout the world, the 1960s were characterized by the free exchange of information and international cooperation. In 1960, the International Atomic Energy Agency established the journal *Nuclear Fusion*. The same organization also sponsored a conference on controlled thermonuclear research in Salzburg in September 1961.[77]

Hayakawa first became a member of the management committee of the Institute of Plasma Physics in late 1961. Shortly thereafter, he was made a member of the National Universities Research Institute Council for a two-year term, enabling him to pursue his goal of building up research infrastructure within a nationwide context. The Science Council Special Committee for Nuclear Fusion continued its activities until June 25, 1966, after which it was incorporated as a sub-committee of the Special Committee for Nuclear Power.[78]

Over the next decades, enthusiasm for fusion would wax and wane. The mid-1960s was a time of low regard for plasma and fusion physics. By the late 1960s, the performance of the stellarator had improved, providing some glimmer of hope for researchers. It was, however, the Russian variation on the "pinch," what is called the "tokamak," that caused renewed interest in 1968. The tokamak was able to sustain a high temperature and was not unlike a reactor plasma. Furthermore, the tokamak did not suffer from the microinstabilities that were causing problems with the stellarator. The 1960s thus ended on a high note for fusion. The early 1970s would be a time during which hopes were centered upon the tokamak as a means of helping to realize a fusion reactor, an alternative, possibly "clean" source of energy.[79]

III   HAYAKAWA AND NAKASONE IN SPACE

In the 1950s, cosmic ray physicists such as Hayakawa developed balloon launching for observations. In the 1960s, space overtook atomic energy as the most exciting technological area and Japan too did not want to be left behind. Although Hayakawa did not play a leading role in space development, he was certainly very much involved. Balloon launchings were incorporated into the activities of the ISAS in the early 1960s. As a result of the increased availability of sounding rockets and balloons courtesy of the Institute, the decade saw a great rise in research in space astronomy. In the mid-1960s, Hayakawa led a group at Nagoya University, which performed observations of x-rays, ultraviolet radiation, and infrared radiation. Rocket observations were superseded by the use of artificial scientific satellites. The 1970s saw astrophysicists and ISAS become involved in satellite launching, beginning with "Hakuchō" (1979), "Hinotori" (1981), "Tenma" (1983), and "Ginga" (1987). Hayakawa examined from a theoretical point of view, the mechanism of x-ray emission of neutron stars and black holes by looking at such experimental data. He was also active in the area of infrared astronomy, using balloon observations to obtain a profile of the Milky Way and the evolution of the galaxy. In 1981, he won the Asahi Prize for the second time, as part of a team for x-ray research using the satellite Hakuchō and was made a member of England's Royal Astronomical Society in 1983.[80] In the 1980s, x-ray astronomers made use of the satellites, and infrared astronomers enthusiastically used balloons and rockets.[81] As in the case of the nuclear power industry, the aerospace industry was late in developing due to Occupation prohibitions on aviation and military-related research and development.[82]

On September 8, 1951, a peace treaty was signed in San Francisco, which was to be effective from April 28, 1952. Many of the limitations that had been placed on the Japanese were lifted, and this included the prohibitions on atomic energy and aerospace research. MITI, the Ministries of Transport and Education, and the Defense Agency put together a plan for the recommencement of aeronautical research. A sub-committee was established within the Scientific and Technical Administration Committee (STAC) to handle the matter and, as a result, a fact-finding mission was sent to Europe and the United States in October 1953. In July 1954, an Aeronautical Technology Council was established within the prime minister's office and based upon one of its reports, an Aeronautical Technology Research Institute was established on July 11, 1955, with the largest budget out of any national research institute in the country. The Japan Aeronautics Society commenced in June 1953, and in 1954 and 1955, respectively, departments for aeronautics were reestablished at the Tokyo University and Kyoto University. In July 1954, the Japan Jet Engine Company, which had started the previous year, successfully assembled an engine.[83]

After a report was submitted by the Aeronautical Technology Council, an Aeronautical Technology Research Institute was established in July 1955. This later moved to the site of the former Central Aeronautical Research Institute in Mitaka, Tokyo, in April 1956. The Institute became affiliated with the Science and Technology Agency, which was created the following month. While the initial budget of the Institute was small (17.13 million yen or US$47,583), by 1959, it had reached over 1.5 billion yen (US$4.167 million).

Japan's desire to gain international prestige through participation in the International Geophysical Year (IGY) stimulated space research. In 1954, an Aerospace Committee met in Rome to organize the IGY, a worldwide project to study the earth involving 67 countries and over 10,000 scientists and technicians.[84] A Japanese Committee was formed in the spring of 1955 to coordinate a rocket project to coincide with the IGY. The latter committee was comprised of staff of the Institute of Industrial Science (IIS) at Tokyo University and members from the National IGY Committee.[85] The IGY helped to transform rocket development into astrophysical research. Researchers became less concerned with operational needs and more interested in space science.[86] Postwar rocket development began in Japan in 1955 when experiments were carried out on a "Pencil" rocket by Professor Hideo Itokawa at the University of Tokyo's IIS. During World War II, Itokawa had been an engineer for the Nakajima Aircraft Company.[87]

Pencil rocket experiments conducted at the Institute from March to April 1955 marked the beginning of space observation R&D. These experiments were the first step in the program to complete an observation rocket system for the IGY, which spanned from July 1957 to December 1958. The period 1957–1958 coincided with a time of intense solar activity, which was thought appropriate for the IGY. A Special Committee for the IGY was organized under the auspices of the International Council of Science Unions (ICSU).[88]

Due to the scale of Japan's IGY project, it was decided that the Science Council should establish a Special Committee on Rocket Observation in April 1956, to take over from the IIS's committee. The Special Committee consisted of 11 scientist-members. The chairman was Kankuro Kaneshige (Tokyo University) and he was assisted by three secretaries: Kenichi Maeda (Kyoto University), Hideo Itokawa (Tokyo University), and Takeo Hatanaka (Tokyo University).[89] Although the Institute was responsible for the sounding rocket and observations of atmospheric pressure, other institutions were involved with the collection of scientific data: Institute for Nuclear Study (Tokyo University) would carry out observations by rockoon (balloon-launched rocket); Tokyo Astronomical Observatory (Tokyo University); Geophysical Institute (Tokyo University); Department of Electronics (Kyoto University); Polytechnic Department (Osaka City University); Radio Research Laboratories; Scientific Research Institute; and the Electrical Communication Laboratory [Nippon Telegraph and Telephone (NTT) Public Corporation].[90]

The main body of the Pencil rocket was 1.8 centimeters in diameter, 23 centimeters in length, and weighed 215 grams. The Pencil was the first of many small-scale rockets. It was followed by larger rockets such as Baby, Kappa, Lambda, and Mu. The "Baby" had an outside diameter of 8 centimeters and measured 1 meter, with a booster adding 30 centimeter to that length. The Pencil and Baby rockets were actually the results of R&D conducted by an avionics and supersonic aerodynamics research group at the Institute, which predated the IGY program. The K-6 (Kappa) rocket was a further manifestation of Japan's enthusiasm for rocket observation, the rocket being completed in June 1958. It, too, was relatively small-scale with a range of 60 kilometers.[91]

The IGY ostensibly provided a reason for rocket development, which would enhance national prestige. After the official IGY program had come to an end, Japan's rocket observation budget (for 1959) was cut from 175 million yen (US$486,111) of the previous year (1958) to 85 million yen (US$236,111). A Rocket Observation

Council (Roketto Kansoku Kyōgikai or ROKK) was established, which incorporated the activities of the former Science Council Special Committee for Rocket Observation. ROKK was run by both basic science practitioners and space specialists. The IIS was originally an institute for engineering-based R&D, but with growing interest in space observation, the number of basic scientists participating in institute activities grew.[92]

On October 4, 1957, the USSR succeeded in launching the world's first artificial satellite, Sputnik 1. On November 3, Sputnik 2 was launched complete with dog. These events surprised the world and marked the beginning of the space race between the United States and the USSR. In June 1959, Nakasone became a state minister, serving as both director of the Science and Technology Agency and chairman of the JAEC for one year. In July 1959, upon Nakasone's suggestion, the Agency set up a Preparatory Committee for the Promotion of Space Science and Technology within the Research Planning Bureau. The Committee conducted discussions with scholars on appropriate R&D policy. This led to the establishment in May 1960 of the Space Development Council as an advisory body to the prime minister.[93]

In November 1960, Nakasone was reelected (for the seventh time) with the Hayato Ikeda government. In January 1961, the Cabinet Committee for the Examination of the Constitution selected Nakasone for a fact-finding tour of North, Central, and South America. In August, he became chairman of the LDP Special Committee for Science and Technology. Nakasone thus emerges as a powerful politician, with a strong focus on science and technology, and a distinct liking for international travel. In November 1962, this even took him as far as Antarctica. Physicists feared that such interest might see Japan joining other nations in the race to carve up the last frontier and gain further territory.[94]

Meanwhile, at the Research Institute of Fundamental Physics, Kyoto University, the director Hideki Yukawa had encouraged Chūshirō Hayashi to conduct research in astrophysics and cosmology. Mituo Taketani and Hayakawa organized the first of a series of workshops to discuss astrophysical problems at RIFP in 1956. Research in theoretical astrophysics was stimulated as a result, with interest in observations and experiments a spinoff as well. This led to the establishment of astrophysics groups in physics departments in Hokkaido University, Nagoya University, and Kyoto University. Researchers in space astronomy thus tended to be physicists, the leaders in the field having mainly come from the field of cosmic-ray research.[95]

In 1959, COSPAR (Committee for Space Research) was established by the ICSU. In January of that year, the Science Council set up a Space Research Liaison Committee, which became a Special Committee the following year, and then reverted to former status in 1973. This Committee became the center of discussions concerning the future and the organization of research. In July 1960, Hayakawa became involved with research connected with an observation rocket at the Tokyo University IIS and again the following year in April 1961. For much of the 1960s, Hayakawa would act as a consultant to the Institute and later to the ISAS. In July 1960, he was made a member of the Science Council Special Committee for Space Research, and appointed secretary in October. On May 29, 1962, the Science Council proposed to the government that space science should be promoted and that a space science research institute be established. The government requested the Ministry of Education to establish such an institute and the Ministry, in turn, asked the University of Tokyo to do so. Few other universities in Japan were capable of establishing a research institute of the scale that was required. The existing activities of the IIS provided further good reasons for locating the new institute at Tokyo. The end-result was the establishment of the Tokyo University ISAS (see figure 7.4).[96]

A problem was that in the University, there were already 14 research institutes, 4 of which were joint-use institutes. The

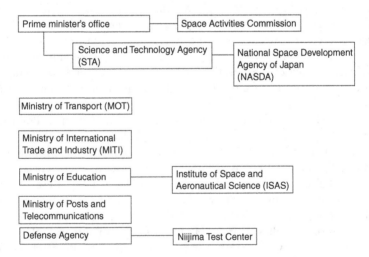

Figure 7.4 Major elements of the organization of government space R&D, ca. 1950s–1960s.

University thus decided to incorporate the Aeronautical Research Institute into the rocket research group of IIS, increase the number of basic science researchers and thus establish an institute. In January 1964, Hayakawa became a member of the preparatory committee for the establishment of ISAS, and later became closely involved in its research program and management of the Institute.[97]

For a long period, the Institute and its successor, the ISAS, was the Japanese space program. The Technical Research and Development Institute of the Japan Defense Agency established the Niijima rocket and missile test center in 1962 (see figure 7.4), but official publications tend not to mention the presence of the Defense Agency in the organization of space research for fear of criticism over possible military applications of such technology.[98]

In January 1963, Nakasone established the League for the Promotion of the Public Election of the prime minister, and was chief of the promotional headquarters for the League's national movement. In November, he was reelected for the eighth time with the Ikeda government and in January 1964, became chairman of the LDP Committee for the Drafting of the 1964 Campaign Direction. Nakasone's interests extended to the arts, with chairmanship of the Diet member's League for the Promotion of the Arts from March 1964. This was in keeping with his self-image for he viewed himself as a writer. In March 1965, Nakasone became chairman of the sub-committee for Asia and Africa within the LDP Foreign Policy Study Group and the next month was appointed an adviser to Ambassador Kawashima who attended the tenth anniversary of the First Asia and Africa Council, which was held in Indonesia.[99]

The Science and Technology Agency that Nakasone had been instrumental in establishing, proceeded with its own space program, which differed substantially from that of the University of Tokyo. The apparent success by Communist China in 1966 in firing a nuclear-capped missile shocked the Japanese into more activity. In 1969, the STA established the National Space Development Agency of Japan (NASDA), with the initial aim of launching a satellite (see figure 7.4). In contrast to Itokawa's group, which insisted on developing technology domestically, NASDA adopted U.S. technology for its Delta rocket and built a three-stage rocket capable of launching a 130-kilogram geostationary satellite. Fifteen years later in 1986, NASDA was able to launch a three-stage rocket, which could place a 550-kilogram satellite into orbit.[100] In 1981, ISAS became a national inter-university research institute and was renamed the Institute of Space and Astronautical Science. While ISAS activities have tended to be more modest in scale compared to NASDA, the former have tended to stress continuity and frequency with their programs, as can be seen in the case of scientific satellites.[101]

There are many government agencies, research institutes, industries, and committees involved in space research but the organization and coordination of research activities have been, as in the case for nuclear power, sadly lacking. Although there has been some overlap in the past with administration in the case of Noboru Takagi when he was both director of the ISAS at the University of Tokyo, and director of the National Space Development Center of the Science and Technology Agency,[102] the three most important government bodies for space research each have their own facilities and programs: the Institute of Space and Astronautical Science (now a national research institute) has the Kagoshima Space Center; the Defense Agency has the Niijima Test Center; and there is also the National Space Development Center on Tanegashima.[103]

Despite the lack of coordination, it was agreed that the Ministry of Education through ISAS would play the leading role in the development of scientific satellites and that ISAS would not develop launching vehicles beyond a certain size. The National Space Development Center of the Science and Technology Agency took the responsibility for developing application satellites and for using launching vehicles larger than those of ISAS.[104]

IV   HAYAKAWA AND ACCELERATORS[105]

In the case of accelerator physics, physicists had a much more difficult time in convincing the government of the need for expensive facilities. The tensions and tradeoffs between scientists and the bureaucracy, which were seen in fusion research and space development, were even more pronounced in the case of accelerators. Hayakawa was again an important actor in this area of physics as well, particularly with respect to his efforts on behalf of the KEK National High Energy Physics Laboratory at Tsukuba, which was established in 1970 under the directorship of Shigeki Suwa, whom Hayakawa had persuaded to return from the United States. The Laboratory centered around the construction of a 10-GeV proton synchrotron on 200 hectares of land within the Tsukuba Science City, outside of Tokyo. This was after 12 years of debate and frustration. The prolonged process showed that the power of the physicists was continuously contested. KEK was the first National Research Institute for Joint Use by Universities. The proposed proton synchrotron was completed in 1976 (achieving a beam of 12 GeV), followed by a photon factory in 1982. These projects were followed by Tristan, a 30-GeV electron–positron collider.[106]

CONCLUSION

Unlike Japan's experience, planning for the Fermi National Accelerator Laboratory in the United States had always emphasized that a strong director was necessary. This contrasted with the Japanese approach, where physicists went to great pains to ensure that a "democratic" management structure would be in place.[107] But in Japan, like in the United States, teacher–student relationships were important. The bond between Tomonaga and Hayakawa worked in a positive way to encourage both to contribute to the establishment of major research facilities. There was, however, a tension between such traditional networks and the desire for a more open scientific community. Also, the elitist tendencies of physics communities to concentrate funding in those most deserving conflicted with the democratic notion of equal distribution of research funding, which the Japanese were attempting to implement. These fundamental problems would greatly delay the construction of Japan's large-scale accelerator.[108]

It did not help that expensive accelerator projects were not considered part of the Japanese atomic energy program, unlike in the United States where national security included leadership in high-energy physics.[109] The Institute for Nuclear Study in Tokyo and the proposed KEK High Energy Physics Laboratory avoided research into weapons or the generation of nuclear power. Hayakawa and other physicists pointed to the prestige that would be derived from a world-class facility.

We also saw earlier in this chapter how Hayakawa and other physicists "won" the debate over "Plan A" and "Plan B" for nuclear fusion. A basic science approach was adopted, which fostered the interests of university-based researchers. It was a pyrrhic victory in the sense that the government was clearly aware of the greater likelihood of commercial benefits being derived from controlled nuclear fission rather than nuclear fusion. The Japan Atomic Energy Research Institute, which tended to represent the establishment of government–business interests, would become involved in plasma research when it suited them, later in 1968, as part of a special national research project.[110]

Space research would also be basically left to physicists and engineers until the Sputnik boom, when it was at the behest of the right wing STA director Yasuhiro Nakasone that the government and the private sector became more heavily involved. The coexistence of both the Ministry of Education backed ISAS and the

Science and Technology Agency supported NASDA shows that in space development, university-based physicists and bureaucracy have come to a mutually agreed standoff in which the former concentrates on satellites of a more scientific nature, and the latter tends to pursue activities with greater commercial potential.

The Science Council of Japan was established in order to maximize the interests of scholars from diverse fields by collective representation via a democratic decision-making structure. The array of councils, committees, and commissions reflected how the powerful did not hold power individually but, rather, through organizations. The scientists clung to the notion of a democratic consensus in decision-making in order to maximize what power they were left with and to further their own particular interests. Physicists such as Hayakawa could wield considerable power within the scientific community, their own milieu, but without adequate political influence, their science-led vision for Japan could not be fully realized.

CONCLUSION

With Japan's defeat in the Pacific War, people spoke of the need to build a New Japan. While there was widespread horror at the devastation caused by the dropping of the atomic bombs on Hiroshima and Nagasaki, there was also awe at the harnessing of science to cause such destruction. Whereas conservative politicians such as Yasuhiro Nakasone yearned for the past and continued to embrace the idea of samurai spirit, the physicists (some of whom were of samurai background) chose the imported discourses of science, Marxism, democracy, and freedom to frame themselves in a new light. There was championing of the need to think like a scientist, to value freedom of inquiry and expression, and to be skeptical and critical of authority.[1] Instead of portraying knowledge as the domain of a select elite, scientists sought to work toward the democratization of knowledge.[2]

On November 20, 1949, Yoshio Nishina (vice president of the Science Council of Japan), the chemist Naoto Kameyama (president), and Mituo Taketani (councilor) were called to attend the meeting of a special House of Representative committee, and answer questions about the award of the Nobel Prize to Hideki Yukawa who was then still in the United States. When asked what sort of a person Yukawa was, Nishina replied "Shinshi de aru to omoimasu" ("He is a gentleman").[3] This and other statements were reassuring to the committee who saw Yukawa's credibility vouched for by the kind of man he was. And even in Japan, there was a perception that successful men of science were leisured gentlemen.

Taketani, whose appointment as lecturer at Nagoya University had now ended, explained how he, himself, was currently unemployed and relied on writing for a living. This, arguably, confirmed to the committee that he was in quite a different category from Yukawa. A committee member asked Taketani how it was that Yukawa was able to accomplish such important work, given the lack of government financial support. Taketani replied that Yukawa's family was wealthy and that, as a result, was able to pursue research in a "nobi nobi" ("unfettered") manner, free of constraints. Taketani pointed out that Sakata, who assisted Yukawa, was also from a relatively well-to-do family.

Questioning then turned to what attitude Yukawa took with respect to the war. Taketani replied that Yukawa was neither a strong advocate of the war, nor did he actively oppose it during the war. Even in that regard, Taketani suggested, Yukawa was a gentle person. Although Taketani and his friends criticized the war, Yukawa would simply acknowledge their comments rather than agree or disagree with either of them. As for nationalistic spirit, Taketani acknowledged that Yukawa had written a poem, but that it was not so much an expression of active advocacy of the war but simply what an ordinary, sensible person might write. A picture emerges of Yukawa as a dignified man of science, not prone to rash statements or actions, unlike Taketani!

By the late 1950s, there was a perception that Japan had entered a new stage in its history. The postwar period had ended and Japan, it seemed, was converging with the West in terms of modernization and prosperity. In this period of renewed confidence, there was a reappraisal of the role of tradition in Japan's modernization and a feeling that perhaps some premodern values and social forms had been useful.[4] Yukawa turned to Japan's past in order to understand what made Japanese different,[5] and the image of a kimono-clad Tomonaga became more common.

The physicists discussed in this book had a sense of public responsibility as scientists. Their background (often samurai), upbringing and culture, individual character, institutional context, and membership of an international scientific community, all helped shape this. But we also need to acknowledge the special nature of the times in which they lived. They came to prominence as public men in the shadow of the bomb,[6] and the sense of power and obligation that that engendered should not be underestimated.

In January 1951, Tomonaga was interviewed at length by the novelist Jiro Osaragi on the atomic age. The dialogue was reported in the newspaper *Mainichi shinbun*. Tomonaga suggested to Osaragi that

> I often think that physicists, including myself, have double personalities. We draw a line between the cases in which we pursue physics as business and those in which we pursue it for enjoyment. In conducting our business, we have in the past stood aloof from reality on the whole.

However, because of the creation of the atomic bomb, we have become unable to stay aloof any more. One reason why I began to study physics was because I thought that I could study it very freely, in the absence of any restrictions.[7]

This statement reflects the tension between public and private life. Like the samurai before them, elite physicists such as Tomonaga saw their public lives as a formal obligation. Interactions with politicians and bureaucrats such as Nakasone, outlined in this book, were a necessary part of the business of physics. The physicist, in private, sought a form of transcendence by pursuing a field of science that was particularly anchored in universal principles and deemed more fundamental than other areas of the physical sciences.[8]

Both Tomonaga and Yukawa lived extremely public lives, fashioning public personas out of what they had initially hoped would be a private passion for science. Japan's cultural past provided them and others with resources from which to shape their sense of self, and build a new Japan that would never consider possessing or producing nuclear weapons. However, Nakasone never forgot seeing the mushroom cloud over Hiroshima on August 6, 1945. "Since the experience of Hiroshima, nuclear power had been stuck in my mind."[9] Ever the warrior, in 1970, while Director General of the Defense Agency, Nakasone commissioned a research team to conduct a secret study of the cost and time necessary for Japan to develop nuclear weapons. The study concluded that Japan was capable of developing such weapons in five years at a cost of 200 billion yen (US$570 million at the time), but the difficulty in securing testing grounds rendered such a plan unfeasible.[10]

Nakasone would go on to serve as prime minister between 1982 and 1987. He would seek to leave his stamp on Japan, and unlike many prime ministers before him, his vision included science. While allegedly opposed to Japan developing nuclear weapons, he felt that if the United States ever withdrew its nuclear umbrella, Japan would have to explore such a possibility.[11] There are parallels between Japan's modernization in the late nineteenth century and the transformations that Japan underwent in the post–World War II era. In both instances, there was a reliance on Western knowledge and men of talent who were willing to contribute to the public good.

NOTES

CHAPTER 1   THE MAKING OF THE JAPANESE PHYSICIST

1. Inazō Nitobe, *Bushido: The Soul of Japan* (Tokyo: Kodansha International, 2002), p. 152. First published in 1900 by The Leeds and Biddle Co., Philadelphia. Revised edition published in 1905.
2. Nitobe, *Bushido*, p. 134.
3. For a fuller exploration of this, see Kam Louie and Morris Low (eds), *Asian Masculinities: The Meaning and Practice of Manhood in China and Japan* (London: Routledge Curzon, 2003).
4. Motoko Kuwahara, "Japanese Women in Science and Technology," *Minerva*, vol. 39 (2001), 203–216, esp. p. 208.
5. Iwan Rhys Morus, *Michael Faraday and the Electrical Century* (Cambridge, UK: Icon Books, 2004), pp. 7–8.
6. Ray A. Moore, "Adoption and Samurai Mobility in Tokugawa Japan," *The Journal of Asian Studies*, vol. 29, no. 3 (May 1970), 617–32.
7. Sharon Traweek, "Generating High Energy Physics in Japan: Moral Imperatives of a Future Pluperfect," unpublished paper. A version published in David Kaiser (ed.), *Pedagogy and Practice in Physics: Historical and Contemporary Perspectives* (Cambridge, MA: MIT Press, 2005).
8. James Richard Bartholomew, "The Acculturation of Science in Japan: Kitasato Shibasaburo and the Japanese Bacteriological Community, 1885–1920," unpublished PhD dissertation, Department of History, Stanford University, 1971, p. 228.
9. Bartholomew, "Acculturation of Science," p. 234.
10. Thomas W. Burkmann, "Nationalist Actors in the Internationalist Theatre: Nitobe Inazō and Ishii Kikujirō and the League of Nations," in Dick Stegewerns (ed.), *Nationalism and Internationalism in Imperial Japan: Autonomy, Asian Brotherhood, or World Citizenship* (London: Routledge Curzon, 2003), pp. 89–113.
11. Masayoshi Sugimoto and David L. Swain, *Science and Culture in Traditional Japan* (Rutland, Vermont/Tokyo: Tuttle, 1989), pp. 224, 229.

12. Shigeru Nakayama, *Characteristics of Scientific Development in Japan* (New Delhi: The Centre for the Study of Science, Technology and Development, CSIR, 1977), pp. 8–9.

13. Thomas Sprat, "Sprat's Criticisms of Artisans, 1667," in Hugh Kearney (ed.), *Origins of the Scientific Revolution* (London: Longman, 1964), pp. 141–42, esp. p. 141. Extract from his *History of the Royal Society* (1667).

14. James Richard Bartholomew, *The Formation of Science in Japan: Building a Research Tradition* (New Haven: Yale University Press, 1989), pp. 18, 22–23.

15. Bartholomew, *Formation of Science*, p. 21.

16. Bartholomew, *Formation of Science*, pp. 23, 32–34. Quote (from Yukichi Fukuzawa) on p. 34 taken from Masao Maruyama, "Fukuzawa Yukichi no jukyō hihan" ("Yukichi Fukuzawa's Criticism of Confucianism"), in Tokyo Imperial University (ed.), *Tokyo Teikoku Daigaku gakujutsu taikan: Hōgakubu keizai gakubu* (*Tokyo Imperial University Research: Law and Economics*) (Tokyo: Tokyo Imperial University, 1942), p. 415.

17. Bartholomew, *Formation of Science*, p. 4.

18. Andrew E. Barshay, *State and Intellectual in Imperial Japan: The Public Man in Crisis* (Berkeley: University of California Press, 1988), pp. xiii, 24.

19. See Shigeru Nakayama, "Japanese Scientific Thought," in Charles C. Gillispie (ed.), *Dictionary of Scientific Biography: Volume 15, Supplement 1, Topical Essays* (New York: Charles Scribner's Sons, 1978), pp. 728–58.

20. Sugimoto and Swain, *Science and Culture*, pp. 305–306.

21. Bartholomew, *Formation of Science*, p. 133; Shigeru Nakayama, *Teikoku daigaku no tanjō* (*The Birth of Imperial Universities*) (Tokyo: Chūō Kōronsha, 1978), esp. pp. 34–71. For details of this process, see D.E. Westney, *Imitation and Innovation: The Transfer of Western Organizational Patterns to Meiji Japan* (Cambridge, MA: Harvard University Press, 1987).

22. Shigeru Nakayama, trans. Jerry Dusenbury, *Academic and Scientific Traditions in China, Japan and the West* (Tokyo: University of Tokyo Press, 1984), pp. 219–20.

23. Bartholomew, *Formation of Science*, p. 144.

24. James Richard Bartholomew, "The Japanese Scientific Community in Formation, 1870–1920: Part 1," *Journal of Asian Affairs*, vol. 5, no. 1 (1980), pp. 62–84.

25. Kenkichirō Koizumi, "The Emergence of Japan's First Physicists: 1868–1900," *Historical Studies in the Physical Sciences*, vol. 6 (1975), pp. 3–108, esp. pp. 50–51.

26. Nakayama, *Academic and Scientific Traditions*, p. 209.

27. See Nakayama, "Japanese Scientific Thought," pp. 728–58.

28. Nakayama, *Academic and Scientific Traditions*, pp. 210–11.

29. Bartholomew, *Formation of Science*, p. 65.
30. W.H. Brock, "The Japanese Connexion: Engineering in Tokyo, London, and Glasgow at the End of the Nineteenth Century," *British Journal for the History of Science*, vol. 14, no. 48 (1981), pp. 227–43.
31. Nobuhiro Miyoshi, *Meiji no enjinia kyōiku* (*Meiji Engineering Education*) (Tokyo: Chūō Kōronsha, 1983), pp. 195–96.
32. Eikoh Shimao, "Some Aspects of Japanese Science, 1868–1945," *Annals of Science*, vol. 46 (1989), pp. 69–91, esp. pp. 74–75.
33. Hiro Tawara, *Pioneers of Physics in the Early Days of Japan* (Amsterdam: North Holland, ca. 1989), pp. 57–58.
34. Kenkichirō Koizumi, "The Development of Physics in Meiji Japan: 1868–1912," unpublished PhD dissertation, University of Pennsylvania, 1973, pp. 244–47.
35. Erwin Baelz, *Das Leben eines deutschen Arztes im erwachenden Japan* (Stuttgart: J. Engelhorns Nachfolger, 1931). Cited in Masao Watanabe, trans. Otto Theodor Benfey, *The Japanese and Western Science* (Philadelphia: University of Pennsylvania Press, 1990), p. 125.
36. Watanabe, *The Japanese*, p. 5.
37. Cited in Watanabe, *The Japanese*, p. 20.
38. Susumu Tonegawa, "Questioning Japanese Creativity," in *Japanese, Kagaku Asahi* (August 1987), pp. 51–75, cited in Watanabe, *The Japanese*, pp. 130–31.
39. Nakayama, *Academic and Scientific Traditions*, p. 51. For carpentry, see William H. Coaldrake, *The Way of the Carpenter: Tools and Japanese Architecture* (New York: Weatherhill, 1990).
40. For details, see Chikayoshi Kamatani, "History of Research Organization in Japan," *Japanese Studies in the History of Science*, no. 2 (1963), pp. 1–77, esp. pp. 34, 38.
41. According to Bartholomew, this degree was invented in the late 1880s by the Japanese Minister of Education and given for significant contributions to one's field, as judged by one's peers. There are various types of doctorates depending on the field. The *rigaku hakasegō* is for science and is often written as DSc to distinguish it from others. Although there is no exact equivalent elsewhere, it tends toward a German PhD with formal dissertation. For an explanation, see Bartholomew, *Formation of Science*, pp. 51–52.
42. Eri Yagi, "The Statistical Analysis of the Growth of Physics in Japan," in Shigeru Nakayama, David L. Swain, and Eri Yagi (eds), *Science and Society in Modern Japan: Selected Historical Sources* (Tokyo: University of Tokyo Press, 1974), pp. 108–113.
43. For further details of the Imperial Institute, see David Cahan, *An Institute for an Empire: The Physikalisch-Technische Reichsanstalt, 1871–1918* (Cambridge: Cambridge University Press, 1989).
44. See Michael A. Cusumano, " 'Scientific Industry': Strategy, Technology, and Entrepreneurship in Prewar Japan," in William D. Wray (ed.), *Managing Industrial Enterprise: Cases from Japan's*

Prewar Experience (Cambridge, MA: Council on East Asian Studies, Harvard University, 1989), pp. 269–315, esp. p. 274. Quoted passage in Kiyonobu Itakura, Kiyonobu, and Eri Yagi, "The Japanese Research System and the Establishment of the Institute of Physical and Chemical Research," in Nakayama, Swain, and Yagi (eds), *Science and Society*, pp. 158–201, esp. pp. 191–92.

45. Shimao, "Some Aspects," pp. 69–91, esp. pp. 84–85.
46. Tetu Hirosige, *Kindai kagaku saikō* (*Reevaluating Modern Science*) (Tokyo: Asahi Shinbunsha, 1979), pp. 181–83.
47. Shimao, "Some Aspects," p. 86.
48. See Table 115, "Estimated Trained Personnel in Selected Scientific and Technical Professions, 1947," in General Headquarters, Supreme Commander of the Allied Powers, *Japanese Natural Resources: A Comprehensive Survey* (Tokyo, 1949), p. 519.
49. General Headquarters, p. 506.
50. Shigeru Nakayama, *Science, Technology and Society in Postwar Japan* (London: Kegan Paul International, 1991), pp. 61–62.
51. Tetu Hirosige, *Kagaku no shakaishi: Kindai Nihon no kagaku taisei* (*The Social History of Science: The Organization of Science in Modern Japan*) (Tokyo: Chūō Kōronsha, 1973), p. 295.
52. David Kaiser, "Cold War Requisitions, Scientific Manpower, and the Production of American Physicists after World War II," *Historical Studies in the Physical and Biological Sciences*, vol. 33, part 1 (Fall 2002), pp. 131–59.
53. Barshay, *State and Intellectual*, p. 228.
54. Chalmers Johnson, *MITI and the Japanese Miracle: The Growth of Industrial Policy, 1925–1975* (Stanford: Stanford University Press, 1982), p. 307.

Chapter 2   Mobilizing Science in World War II: Yoshio Nishina

1. Letter, May 5, 1933, Y. Nishina to G. Hevesy, in *G. Hevesy–Y. Nishina: Correspondence, 1928–1949* (Tokyo: Nishina Memorial Foundation, 1983), pp. 10–11.
2. He does, however, lend his name to the "Klein–Nishina formula," which he and O. Klein arrived at in 1928 to describe the differential cross-section of Compton scattering, enabling one to calculate the probability or frequency of scattering of high-energy quanta by electrons. See Yūji Yazaki, "Klein-Nishina kōshiki dōshutsu no katei (1): Riken no Nishina shiryō o chūshin ni" ("How was the Klein-Nishina Formula Derived? (Part 1): Based on the Nishina Archives at Riken"), *Kagakushi kenkyū* (*Studies in the History of Science*), series 2, vol. 31, no. 182 (Summer 1992), pp. 81–91.
3. Joseph M. Goedertier, *A Dictionary of Japanese History* (New York: Walker/Weatherhill, 1968), p. 35.

4. Fumio Yamazaki, "A Short Biography of Dr. Yoshio Nishina," in Sin-itirō Tomonaga and Hidehiko Tamaki (eds), *Nishina Yoshio: Denki to kaisō (Yoshio Nishina: Biography and Reminiscences)* (Tokyo: Misuzu Shobō, 1952), pp. 3–15. Also, Tetsuo Tsuji, "The Life of Yoshio Nishina" (in Japanese), *Nihon Butsuri Gakkaishi*, vol. 45, no. 10 (October 1990), pp. 712–19.

5. In 1939, Shibaura Engineering Works (Shibaura Seisakujo) merged with Tokyo Electric (Tokyo Denki) to form Tōshiba, the largest electrical concern in Japan. Andrew Gordon, *The Evolution of Labor Relations in Japan: Heavy Industry, 1853–1955* (Cambridge, MA: Council on East Asian Studies, Harvard University, 1988), pp. 11–12. Also see Chikayoshi Kamatani, "The Role Played by the Industrial World in the Progress of Japanese Science and Technology," *Cahiers d'histoire mondiale*, vol. 9, no. 2 (1965), pp. 400–421.

6. See, e.g., the career of Ken'ichi Fukui of Kyoto University, joint winner in 1981 of the Nobel Prize for chemistry, described in James R. Bartholomew, "Perspectives on Science and Technology in Japan: The Career of Fukui Ken'ichi," *Historia Scientiarum*, vol. 4, no. 1 (1994), pp. 47–53. Such interdisciplinarity was not confined to Japan. The Nobel Prize–winning physicist Paul Dirac, like Nishina, graduated in Electrical Engineering. See Laurie M. Brown and Helmut Rechenberg, "Paul Dirac and Werner Heisenberg: A Partnership in Science," in Behram N. Kursunoglu and Eugene P. Wigner (eds), *Reminiscences about a Great Physicist: Paul Adrien Maurice Dirac* (Cambridge: Cambridge University Press, 1987), pp. 117–61, esp. p. 137.

7. Rene Taton (ed.), trans. A.J. Pomerans, *Science in the Twentieth Century* (London: Thames and Hudson, 1966), p. 208.

8. These included: Hendrik A. Kramers (Netherlands: 1916–1926), Werner Heisenberg (Germany: 1924–1925, 1926–1927), Wolfgang Pauli (Switzerland: 1922–1923), E. Pascual Jordan (Germany: 1927), Oskar Klein (Sweden: 1918–1922; 1926–1931), Paul A.M. Dirac (England: 1926–1927), Charles G. Darwin (England: 1927), Ralph H. Fowler (England: 1925), Samuel A. Goudsmit (Netherlands: 1926, 1927), and George Gamow (Russia: 1928–1929, 1930–1931). The information in the brackets gives country of origin and time of visit.

9. Taton, *Science in the Twentieth Century*, pp. 203–204.

10. Yoshio Nishina, "On the L-absorption spectra of the elements from $Sn(50)$ to $W(74)$ and their relation to the atomic constitution," *The Philosophical Magazine*, vol. 49 (1925), pp. 521–37. This paper was unusual for him in that he was the only author. Nishina would use this as one of the papers in his submission for his DSc, which he received in 1930.

11. For a list of visitors to the Institute for Theoretical Physics, see Peter Robertson, *The Early Years: The Early Years: The Niels Bohr Institute, 1921–1930* (Copenhagen: Akademisk Forlag, 1979), pp. 156–59.

12. Yamazaki, "A Short Biography," p. 6.

13. See Robertson, *The Early Years*, pp. 49–50, 156–59.

14. During Nishina's visit to Paris, he received news from Bohr and a further check from the Rask-Ørsted Foundation to cover some of the expenses he incurred during his longer-than-expected stay in Copenhagen. Nishina had originally received a stipend from the Foundation, but this had run out. Letter, September 1, 1927, Y. Nishina to N. Bohr, in *Y. Nishina's Letters to N. Bohr, G. Hevesy and Others, 1923–1928* (Tokyo: Nishina Memorial Foundation, 1985), pp. 16–17.

15. For details of his schedule, see Letter, October 26, 1928, Nishina to Prof. and Mrs. Hevesy, in *Y. Nishina's Letters to N. Bohr, G. Hevesy and Others*, pp. 24–25.

16. One such visit was in mid-August 1925. Letters, August 12, September 17, 1925, Nishina to Hevesy, in *Y. Nishina's Letters to N. Bohr, G. Hevesy and Others*, pp. 12–15.

17. Letter, October 31, 1928, Nishina to Bohr, in *Y. Nishina's Letters to N. Bohr, G. Hevesy and Others*, pp. 25–26.

18. Letter, June 11, 1929, Y. Nishina to P.A.M. Dirac, in *P.A.M. Dirac-Y. Nishina Correspondence, 1928–1948* (Tokyo: Nishina Memorial Foundation, 1990), pp. 5–6. The visit of Dirac and Heisenberg to Japan is discussed in Brown and Rechenberg, "Paul Dirac and Werner Heisenberg," pp. 132–41.

19. Yamazaki, "A Short Biography," pp. 7–8. Also Shimpei Miyata, *Kagakushatachi no jiyū na rakuen: Kōei no Rikagaku Kenkyūjo (The Scientists' Paradise: The Famous Institute of Physical and Chemical Research)* (Tokyo: Bungei Shunjū, 1983), p. 186.

20. Letters, November 25, 1929 and November 6, 1932, Nishina to Bohr, in *Y. Nishina's Correspondence with N. Bohr and Copenhageners, 1928–1949* (Tokyo: Nishina Memorial Foundation, 1984), pp. 13–14, 19–20; Letter, July 18, 1933, Nishina to Hevesy, in *G. Hevesy-Y. Nishina: Correspondence, 1928–1949* (Tokyo: Nishina Memorial Foundation, 1983), pp. 13–14.

21. "Nishina kenkyūshitsu no hassoku" ("The Establishment of the Nishina Laboratory," *Mugendai*, no. 85 (autumn 1990), p. 46.

22. "Nishina sensei o shinonde" ("Recollections of Prof. Nishina"), round-table discussion, *Shizen* (March 1971), pp. 30–42.

23. Letter, November 6, 1932, Nishina to Bohr, in *Y. Nishina's Correspondence with N. Bohr and Copenhageners*, pp. 19–20.

24. Invention of the Cavendish Laboratory, Cambridge University.

25. The "cyclotron" was formally referred to by Lawrence as "magnetic resonance accelerator." See J.L. Heilbron and Robert W. Seidel, *Lawrence and his Laboratory: A History of the Lawrence Berkeley Laboratory, Volume 1* (Berkeley: University of California Press, 1989), p. 84.

26. For an easy-to-understand discussion of the cyclotron, see the following: Henry A. Boorse, Lloyd Motz, and Jefferson Hane Weaver, *The Atomic Scientists: A Biographical History* (New York: John Wiley

and Sons, 1989), pp. 358–59; Robert R. Wilson and Raphael Littauer, *Accelerators: Machines of Nuclear Physics* (Garden City, New York: Anchor Books, 1960), pp. 87–101.

27. Letter, July 18, 1933, Nishina to Betty Schultz, in *Y. Nishina's Correspondence with N. Bohr and Copenhageners*, pp. 26–28.

28. Letter, May 5, 1933, Nishina to Hevesy, in *G. Hevesy–Y. Nishina: Correspondence, 1928–1949*, pp. 10–11.

29. Named after John Douglas Cockcroft and Ernest Walton. It is interesting to note that Robert Jemison Van de Graaff (of Van de Graaff generator fame), Cockcroft, and Nishina all started their careers as engineers. Heilbron and Seidel, *Lawrence and his Laboratory*, p. 66.

30. Sin-itirō Tomonaga, "Dr. Yoshio Nishina, His Sixtieth Birthday," in *In Memory of the Late Dr. Yoshio Nishina*, pp. 2–3—a six page booklet possibly originally inserted in an issue of *Progress of Theoretical Physics* (1951), Sakata Archival Library, Nagoya University.

31. Yuichirō Nishina, "Chichi no kaisō" ("Recollections of my Father"), *Nihon Butsuri Gakkaishi*, vol. 45, no. 10 (October 1990), pp. 724–26.

32. Daniel J. Kevles, *The Physicists: The History of a Scientific Community in Modern America* (New York: Vintage Books, 1979), p. 272.

33. In a letter to the Bohrs, he wrote of how "the accumulation of every day task [*sic*] in our Institute prevents me from writing any letters except those for urgent matters. It is a very sad condition which, however, I am unable to change at present." Letter, January 8, 1936, Nishina to Prof. and Mrs. Bohr, in *Y. Nishina's Correspondence with N. Bohr and Copenhageners*, pp. 52–54.

34. For details of the assistance provided by Lawrence, see Charles Weiner, "Cyclotrons and Internationalism: Japan, Denmark and the United States, 1935–1945," *14th International Congress on the History of Science Proceedings: Volume 2*, held in Tokyo and Kyoto, Japan 1974 (1975) pp. 353–65. Also Heilbron and Seidel, *Lawrence and his Laboratory*, pp. 318–20, 520.

35. Kiyonobu Itakura, Tōsaku Kimura, and Eri Yagi, *Nagaoka Hantarō den* (*A Biography of Hantarō Nagaoka*) (Tokyo: Asahi Shinbunsha, 1973), pp. 636–37.

36. Cited in Guenther Stein, *Made in Japan* (London: Methuen and Co., 1935), pp. 197–98. Also referred to in Tetu Hirosige, *Kagaku no shakaishi: Kindai Nihon no kagaku taisei* (*The Social History of Science: The Organization of Science in Modern Japan*) (Tokyo: Chūō Kōronsha, 1973), p. 135.

37. Letters, September 7, 1934 and August 3, 1936, Nishina to Bohr, in *Y. Nishina's Correspondence with N. Bohr and Copenhageners*, pp. 42–43, 56–58. It had been hoped that Bohr would have been be able to visit Japan much earlier, as far back as 1930, but work and the death of Bohr's son Christian in a boating accident in mid-1934 put such travel plans on hold. See Letter, April 6, 1929, Nishina to Bohr, in *Y. Nishina's Correspondence with N. Bohr and Copenhageners*, pp. 2–3.

38. *Tōkyō asahi shinbun*, April 7, 1937. Translation of article by Tetu Hirosige, cited in Weiner, "Cyclotrons and Internationalism," pp. 356–57.

39. Letter, November 14, 1936, Ryōkichi Sagane to Ernest O. Lawrence, Ernest Orlando Lawrence Correspondence and Papers, BANC MSS 72/117C, Carton 9, Folder 39, The Bancroft Library, University of California, Berkeley.

40. See, e.g., Eizō Tajima, "Riken no saikurotoron" ("Riken's Cyclotron"), in Hidehiko Tamaki and Hiroshi Ezawa (eds), *Nishina Yoshio: Nihon no genshi kagaku no akebono* (*Yoshio Nishina: The Dawn of Atomic Science in Japan*) (Tokyo: Misuzu Shobō, 1991), pp. 119–27, esp. p. 124.

41. Weiner, "Cyclotrons and Internationalism," p. 358.

42. Letter, February 17, 1937 and reply, March 8, 1937, Nishina to Hevesy, in *G. Hevesy–Y. Nishina: Correspondence, 1928–1949*, pp. 16–17, 17–18.

43. Satoshi Saitō, "The Finances of the Institute of Physical and Chemical Research and the Research Funds," *Nihon Butsuri Gakkaishi*, vol. 45, no. 10 (October 1990), pp. 761–65, esp. p. 761.

44. Otto Hahn, laboratory diary entry, November 24, 1938, cited in Burghard Weiss, "The 'Minerva' Project, The Accelerator Laboratory at the Kaiser Wilhelm Institute/Max Planck Institute of Chemistry: Continuity in Fundamental Research," in M. Renneberg and M. Walker (eds), *Science, Technology and National Socialism* (Cambridge: Cambridge University Press, 1994), pp. 271–90, 400–405 note 19, p. 402. Paper referred to is Y. Nishina, T. Yasaki, K. Kimura, and M. Ikawa, "Artificial Production of Uranium Y from Thorium," *Nature*, vol. 142 (1938), p. 874, letter (dated September 13). Source: Tamaki and Ezawa, *Yoshio Nishina*, p. 311.

45. Sin-itirō Tomonaga, *Tomonaga Sin-itirō chosaku-shū, bekkan 2: Nikki, shokan* (*Collected works of Sin-itiro Tomonaga, Supplementary Volume 2: Diary and Letters*) (Tokyo: Misuzu Shobō, 1985), pp. 145, 155.

46. Committee for Yukawa Hall Archival Library, *YHAL Resources: Hideki Yukawa: (III)* (Kyoto: Research Institute for Fundamental Physics, 1985), EDT 050, pp. 1–2.

47. Yasutaka Tanikawa, "Introduction and Biographical Sketch," in Hideki Yukawa, *Hideki Yukawa: Scientific Works* (Tokyo: Iwanami Shoten, 1979), pp. vii–xx, esp. p. xvii.

48. Shōichi Sakata, *Shoichi Sakata: Scientific Works* (Nagoya: Publication Committee of Scientific Papers of Professor Shōichi Sakata, 1977), pp. viii–ix.

49. Nobel Foundation, *Nobel Lectures: Physics, 1963–1979* (New York: Elsevier, 1972), p. 137.

50. Satio Hayakawa, "The Development of Meson Physics in Japan," in Laurie M. Brown and Lillian Hoddeson (eds), *The Birth of Particle Physics* (Cambridge: Cambridge University Press, 1983), pp. 82–107, esp. p. 102.

51. Letter, August 22, 1940, Ernest O. Lawrence to Yoshio Nishina, Ernest Orlando Lawrence Correspondence and Papers, BANC MSS 72/117C, Carton 9, Folder 38, The Bancroft Library, University of California, Berkeley.
52. Hideomi Tuge (ed.), *Historical Development of Science and Technology in Japan* (Tokyo: Kokusai Bunka Shinkōkai, 1968), p. 165.
53. J.W.M. Chapman, "The Transfer of German Underwater Weapons Technology to Japan, 1919–1976," in Ian Nish and Charles Dunn (eds), *European Studies on Japan* (Tenterden, Kent: Paul Norbury Publications, 1979), pp. 165–71, 341–42.
54. Hans-Joachim Braun, "Technology Transfer under Conditions of War: German Aero-Technology in Japan During the Second World War," in Norman Smith (ed.), *History of Technology*, vol. 11 (1986) (London: Mansell Publishing, 1987), pp. 1–23, esp. p. 6.
55. Mituo Taketani, "Theoretical Study of Cosmic Rays in Japan," paper prepared for *International Symposium on the History of Particle Physics, May 28–31, 1980 at Fermi National Accelerator Laboratory, Illinois*, p. 15.
56. Laurie M. Brown, Rokuo Kawabe, Michiji Konuma, and Zirō Maki (eds), "Elementary Particle Theory in Japan, 1930–1960: Proceedings of the Japan–USA Collaborative Workshops," *Progress of Theoretical Physics Supplement*, no. 105 (1991), special issue, p. 40.
57. Yoshinori Kaneseki, "The Elementary Particle Theory Group," 1950, trans. Shigeru Nakayama, David L. Swain, and Eri Yagi (eds), *Science and Society in Modern Japan* (Tokyo: University of Tokyo Press, 1974), p. 242.
58. Kaneseki, "The Elementary Particle Theory Group," p. 242.
59. T. Miyazima (ed.), *Scientific Papers of Tomonaga, Vol. 1* (Tokyo: Misuzu Shobō, 1971), p. ii.
60. J.E. Greene (ed.), *McGraw-Hill Modern Men of Science* (New York: McGraw-Hill, 1966), p. 483.
61. Kaneseki, "The Elementary Particle Theory Group," p. 242.
62. "Tomonaga Sin-itirō hakushi: Hito to gyōseki" ("Dr. Sin-itirō Tomonaga: The Man and his Achievements"), *Physics Monthly*, vol. 1, no. 4 (October 1979), p. 388.
63. Second Yoshio Nishina/Major General Nobuuji meeting, "Uranium Project Research Meeting," February 2, 1944, original document and trans., P. Wayne Reagan Collection, Kansas City, Mo., USA. Cited in Richard Rhodes, *The Making of the Atomic Bomb* (New York: Simon and Schuster, 1986), p. 581.
64. This account is largely derived from three sources: Tetu Hirosige, *Kagaku to rekishi* (*Science and History*) (Tokyo: Misuzu Shobō, 1965), pp. 169–84; *Kagaku no shakaishi: Kindai Nihon no kagaku taisei* (*The Social History of Science: The Organization of Science in Modern Japan*) (Tokyo: Chūō Kōronsha, 1973), pp. 200–222; Chikayoshi Kamatani, "The History of Research Organization

in Japan," *Japanese Studies in the History of Science*, no. 2 (1963), pp. 1–79.

65. Hitoshi Yoshioka, "Science and Technology at the Turning Point: Founding the Technological State," *Canadian Journal of Political and Social Theory*, vol. 8, no. 3 (Fall 1984), pp. 21–39, esp. p. 26.

66. See Karl T. Compton, "Mission to Tokyo," unpublished, typewritten manuscript, October 8, 1945, p. 20, Institute Archives, MIT [T-N C738], courtesy of Prof. R.W. Home, University of Melbourne.

67. Kamatani, "The History of Research Organization in Japan," pp. 40–41, 48–49.

68. Hirosige, *Science and History*, pp. 169, 175.

69. Hirosige, *Science and History*, p. 178.

70. Hirosige, *Science and History*, p. 178.

71. Hirosige, *Science and History*, p. 175.

72. "Gijutsuin" is sometimes translated as "Board of Technology" or "Board of Technics." For further details, see Masakatsu Yamazaki, "The Mobilization of Science and Technology during the Second World War in Japan: A Historical Study of the Activities of the Technology Board Based Upon the Files of Tadashiro Inoue," *Historia Scientiarum*, vol. 5, no. 2 (1995), pp. 167–81.

73. Kamatani, "The History of Research Organization in Japan," pp. 40–41, 48–49.

74. Hirosige, *The Social History of Science*, p. 207.

75. Hideki Yukawa in *Kyokuchi no sekai* (*The World of the Extremities*) (February 1942). Cited in Takayuki Kan, *Kaku jidai no kagaku gijutsu to shakai no genri* (*Science, Technology and Social Principles in the Nuclear Age*) (Tokyo: Hihyōsha, 1984), p. 101 (my translation).

76. Donald Keene, "The Barren Years: Japanese War Literature," *Monumenta Nipponica*, vol. 33 (spring 1978), pp. 67–112, esp. pp. 67–68. Cited in Richard H. Mitchell, *Censorship in Imperial Japan* (Princeton: Princeton University Press, 1983).

77. Kiyonobu Itakura, Tōsaku Kimura, and Eri Yagi, *Nagaoka Hantarō den* (*A Biography of Hantarō Nagaoka*) (Tokyo: Asahi Shinbunsha, 1973), pp. 635–37.

78. Hirosige, *Science and History*, p. 178.

79. Hirosige, *The Social History of Science*, p. 211.

80. Hirosige, *The Social History of Science*, p. 212.

81. Hirosige, *The Social History of Science*, p. 214.

82. Hirosige, *The Social History of Science*, p. 214.

83. Yomiuri Shinbun, *Shōwa shi no tennō: Genbaku tōka* (*The Emperor and Shōwa History: The Dropping of the Bomb*) (Tokyo: Kadokawa Shoten, 1988), p. 99.

84. Charles Weiner, "Retroactive saber rattling?: A note on nuclear physics in Japan," *Bulletin of the Atomic Scientists* (April 1978), pp. 10–12, esp. pp. 10–11.

85. Spencer Weart, "Secrecy, Simultaneous Discovery and the Theory of Nuclear Reactors," *American Journal of Physics*, vol. 45, no. 11 (November 1977), pp. 1049–60, esp. p. 1055.

86. John W. Dower, "Science, Society, and the Japanese Atomic-Bomb Project During World War Two," *Bulletin of Concerned Asian Scholars*, vol. 10, no. 2 (April–June 1978), pp. 41–54, esp. p. 47; *Japan in War and Peace: Essays on History, Culture and Race* (London: Harper Collins, 1995), pp. 55–100. Also see Robert K. Wilcox, *Japan's Secret War* (New York: William Morrow, 1985).

87. Yomiuri Shinbun, *The Emperor and Shōwa History*, pp. 88–89.

88. Yomiuri Shinbun, *The Emperor and Shōwa History*, pp. 89, 91, 93.

89. Yasunori Matogawa, *Hoshi no ōji sama, uchū o yuku: Oda Minoru kara no messeeji* (*The Prince of the Stars Goes to Space: A Message from Minoru Oda*) (Tokyo: Dōbun Shoin, 1990), p. 34. Bunsaku Arakatsu was born in 1890, graduated from the Department of Physics, Kyoto Imperial University, in 1918, and later studied in England with Ernest Rutherford at Cambridge University. He eventually became professor at Taipei Imperial University and later returned to Kyoto University as a professor, with an affiliation as a research fellow at the Chemical Research Institute. Kagaku Bunka Shinbunsha (ed.), supervised by the Ministry of Education, *Nihon kagaku gijutsusha meikan: Rigaku hen* (*Directory of Contemporary Japanese Scientists and Technologists: Science*) (Tokyo: Kagaku Bunka Shuppansha, 1949), p. 31; "Japan's Hidden Physicists," *Nippon Times*, November 22, 1949, GHQ/SCAP files, ESS (E) 06365, National Diet Library, Tokyo.

90. Captain Itō appears to have obtained a doctorate. He graduated from the Department of Electrical Engineering of Tokyo Imperial University in 1924, and, subsequently, studied at Technische Hochschule, Dresden, Germany, graduating in 1927. US Naval Technical Mission to Japan, "Japanese Land-Based Radar: 'Intelligence Targets Japan' (DNI) of 4 Sept. 1945, Fascicle E-1, Target E-03," p. 13, restricted report, December 1945. Courtesy of Emeritus Prof. Rod Home, University of Melbourne.

91. Deborah Shapley, "Nuclear Weapons History: Japan's Wartime Bomb Projects Revealed," *Science*, vol. 199 (January 13, 1978), pp. 152–57, esp. p. 154; Dower, "Science, Society, and the Japanese Atomic-Bomb Project," p. 47.

92. A detailed account can be found in Yomiuri Shinbun, *Shōwashi no tennō: 4* (*The Emperor and Shōwa History: Volume 4*) (Tokyo: Yomiuri Shinbunsha, 1968), pp. 78–229.

93. History of Science Society of Japan (ed.), *Nihon kagaku gjutsushi taikei, 13: butsurigaku* (*An Outline of the History of Science and Technology in Japan, vol. 13: Physics*) (Tokyo: Daiichi Hōki, 1970), p. 441.

94. Shapley, "Nuclear Weapons History," p. 154; Dower, "Science, Society, and the Japanese Atomic-Bomb Project," pp. 49–50.

95. Report, November 4, 1947, S. Nomura, Scientific Education Bureau, Ministry of Education, to Dr. Kelly, Scientific and Technical Division, Economic and Scientific Section, SCAP, "Brief Description on [*sic*] the Policy to Find Out Possibility of Utilizing Atomic Energy Adopted by the then Japanese Military Authorities in the [*sic*] War Time: Report No. 3," GHQ/SCAP records, ESS (B) 11758, National Diet Library, Tokyo.

96. This figure uses the 1930s exchange rate of four yen per U.S. dollar, given in Al Alletshauser, *The House of Nomura* (London: Bloomsbury, 1990), appendix 6.

97. Report, January 6, 1948, Nomura to Kelly.

98. Dower, "Science, Society, and the Japanese Atomic-Bomb Project," p. 51.

99. James Phinney Baxter III, *Scientists Against Time* (Cambridge, MA: MIT Press, 1968), p. 10.

100. Robert serber, *The Los Alamos Primer: The First Lectures on How to Build an Atomic Bomb* (Berkeley: University of California Press, 1992), p. 93.

101. For example, see Hirosige, *The Social History of Science*, p. 218.

102. David Snell, typed eight-page manuscript, 1946, report for the *Atlantic Constitution*, GHQ/SCAP records, Economic and Scientific Section files, ESS (E) 06379, National Diet Library, Tokyo. An extended account of this has been published. See Wilcox, *Japan's Secret War*.

103. Robert Hechter, "Japanese 'Death Ray' Can Kill Men," *Pacific Stars and Stripes*, October 27, 1945, reproduced in History of Science Society of Japan, *Vol. 13: Physics*, p. 473.

104. History of Science Society of Japan (ed.), *Nihon kagaku gijutsushi taikei, 19: denki gijutsu* (*An Outline of the History of Japanese Science and Technology, vol. 19: Electrical Technology*) (Tokyo: Daiichi Hōki, 1969), p. 359.

105. "Short Survey of Japanese Radar," 3 vols., Air Technical Intelligence Group Report no. 115, vol. 1 (Tokyo, November 20, 1945), p. 22, cited in Jerome B. Cohen, *The Japanese War Economy, 1937–1945* (London: Oxford University Press, 1949), p. 241.

106. Eri Yagi and Derek J. de Solla Price, "Japanese Bomb," *Bulletin of the Atomic Scientists*, vol. 18, no. 9 (November 1962), p. 29.

107. Sakata, *Shoichi Sakata: Scientific Works*, p. ix.

108. Manhattan Engineer District Records, National Archives, Washington, DC, MED 76, cited in Richard Rhodes, "Introduction," in Robert Serber, *The Los Alamos Primer: The First Lectures on How to Build An Atomic Bomb* (Berkeley: University of California Press, 1992), pp. ix–xxi, xviii.

109. Yasuhiro Nakasone, "Toward a Nation of Dynamic Culture and Welfare," *Japan Echo*, vol. 10, no. 1 (spring 1983), pp. 12–18.

110. My translation of citation from *Pentohausu* (*Penthouse*) (May 1983), in Michitaka Kido, *Nakasone no kokka kaizō keikaku hihan* (*Criticism*

of Nakasone's Plans for Rebuilding the Nation) (Tokyo: Senshinsha, 1986), p. 134.

111. Yasuhiro Nakasone, "Kagaku Gijutsu Chō setsuritsu made no omoide" ("Memories up until the Establishment of the Science and Technology Agency"), in Science and Technology Agency, *Kagaku Gijutsu Chō: 30 nen no ayumi* (*The Science and Technology Agency: A 30 Year History*) (Tokyo: Sōzō, 1986), pp. 94–95. Originally printed in Science and Technology Agency, *Kagaku Gijutsu Chō: jūnen shi* (Tokyo: Science and Technology Agency, 1966), pp. 27–28.

112. Motoharu Kimura, *Kaku to tomo ni 50 nen* (*50 Years with Things Nuclear*) (Tokyo: Tsukiji Shokan, 1990), p. 46. Also Nishina Foundation (ed.), *Genshi bakudan* (*The Atomic Bomb*) (Tokyo: Kōfūsha Shoten, 1973), courtesy of Hajime Takeuchi.

113. Yamazaki, "A Short Biography," pp. 13–14.

114. Brown et al. (eds), "Elementary Particle Theory in Japan, 1930–1960," p. 40.

115. For example, see Miyata, *The Scientists' Paradise*, p. 245.

116. Tuge, *Historical Development*, pp. 146, 150.

117. For example, see Yōichi Yamamoto, "Nihon Genbaku no Shinsō" ("The Truth about Japan's Atomic Bomb"), *Daihōrin*, vol. 20, no. 8 (August 1953), pp. 6–40, 34; History of Science Society of Japan, *Vol. 13: Physics*, p. 443.

118. For an account of measures taken by Japan to overcome this problem, see Hirosige, *The Social History of Science*, pp. 204–205.

119. For example, see Nobuo Yamashita, "Ma ni awanakatta Nihon no Genbaku" ("Japan's Uncompleted Atomic Bomb"), *Kaizō*, special number (November 15, 1952), pp. 162–65, esp. p. 164; Mituo Taketani, *Taketani Mituo chosakushū: 2, Genshiryoku to kagakusha* (*Collected Works of Mituo Taketani: Volume 2, Scientists and Atomic Energy*) (Tokyo: Keisō Shobō, 1968), p. 355.

120. For example, see "Nihon no butsurigaku" ("Japanese Physics"), *Shizen*, vol. 18 (January 1963), cited in History of Science Society, *Vol. 13: Physics*, p. 418.

121. Reference is made to this survey in Hirosige, *Science and History*, p. 180. Yamamoto writes that the most frequent reply to the question about the period of greatest intellectual freedom during the past decades was the Pacific War period. See Yamamoto, "The Truth about Japan's Atomic Bomb," p. 19.

122. John W. Dower, "The Useful War," *Daedalus*, vol. 119, no. 3 (Summer 1990), pp. 49–70, esp. p. 56. For details of the numbers of graduates, see Ben-Ami Shillony, *Politics and Culture in Wartime Japan* (Oxford: Clarendon Press, 1991), p. 139.

123. First Yoshio Nishina/Major General Nobuuji meeting, "Uranium Project Research Meeting," July 2, 1943, original document and trans., P. Wayne Reagan Collection, Kansas City, Mo., USA. Cited in Rhodes, *The Making of the Atomic Bomb*, p. 580.

124. For a discussion of the activities of intellectuals during the war, see Shillony, *Politics and Culture in Wartime Japan*, esp. chapter 5. For another account of Nishina's wartime activities, see Kenji Itō, "Values of 'Pure Science': Nishina Yoshio's Wartime Discourse between Nationalism and Physics, 1940–1945," *Historical Studies in the Physical and Biological Sciences*, vol. 33, part 1 (2002), pp. 61–86.

125. Letter, October 4, 1945, Karl T. Compton to President Harry S. Truman, Harry S. Truman Papers, 1-A-21, Makoto Iokibe (ed.), *The Occupation of Japan, Part 2: U.S. and Allied Policy, 1945–1952 Microfiche Collection* (Bethesda, MD: Congressional Information Service/Tokyo: Maruzen, 1989).

126. The anthropologist/historian Sharon Traweek argues this for physicists in general. See her *Beamtimes and Lifetimes: The World of High Energy Physicists* (Cambridge, MA: Harvard University Press, 1988), pp. 1–2.

127. For further evidence of this, see Kenji Ito, "Values of 'Pure Science'," pp. 61–86.

CHAPTER 3   THE IMPACT OF
THE ALLIED OCCUPATION: NISHINA
AND NAKASONE

1. Jean-Jacques Salomon, "Science Policy Studies and the Development of Science Policy," in Ina Spiegel-Rosing and Derek de Solla Price (eds), *Science, Technology and Society: A Cross-Disciplinary Perspective* (London: Sage, 1977), pp. 43–70, esp. p. 48.

2. Kinjirō Okabe, "Kagaku Nihon no saiken" ("The Rebuilding of Scientific Japan"), *Kagaku asahi* (August–September 1945), p. 1.

3. William McGucken, *Scientists, Society, and State: The Social Relations of Science Movement in Great Britain, 1931–1947* (Columbus: Ohio State University Press, 1984), p. 307.

4. Historically, the years after the war have been deemed a period of "demilitarization" and "demobilization." This chapter seeks to reconsider this, drawing on SCAP (Supreme Commander for the Allied Powers) records available in the National Diet Library, Tokyo.

5. J.A.A. Stockwin, "The Occupation: Continuity or Change?," *Asian and African Studies* (Israel), vol. 18 (1984), pp. 27–40, esp. p. 35.

6. On August 6, 1945, an atomic (uranium) bomb was dropped on Hiroshima, and on August 9, an atomic (plutonium) bomb was dropped on Nagasaki.

7. They were, however, prohibited from making public their reports. By the end of 1946, around 80 reports had been compiled and submitted to SCAP. After drawn-out deliberations, some of the manuscripts were released in January 1949. It wasn't until 1953, after the Occupation, that a collection of reports by the NRC group were published in Japanese. See Mitutomo Yuasa, *Kagakushi*

(*The History of Science*) (Tokyo: Tōyō Keizai Shinpōsha, 1961), pp. 294–95; and Committee for the Compilation of Materials on Damage Caused by the Atomic Bombs in Hiroshima and Nagasaki, trans. Eisei Ishikawa and David L. Swain, *The Impact of the A-bomb: Hiroshima and Nagasaki, 1945–85* (Tokyo: Iwanami Shoten, 1985), p. 199. Also, Monica Braw, *The Atomic Bomb Suppressed: American Censorship in Japan, 1945–1949*, Lund Studies in International History No. 23 (Malmo, Sweden: Liber Forlag, 1986), pp. 121, 127–28.

8. For a discussion of this, see Theodore Cohen, edited by Herbert Passin, *Remaking Japan: The American Occupation as New Deal* (New York: Free Press, 1987), pp. 7–9.

9. Meirion Harries and Susie Harries, *Sheathing the Sword: The Demilitarisation of Japan* (London: Hamish Hamilton, 1987), p. 21.

10. Chalmers Johnson, *MITI and the Japanese Miracle: The Growth of Industrial Policy, 1925–1975* (Stanford: Stanford University Press, 1982), p. 172. See also, "Japan: Who Governs? An Essay on Official Bureaucracy," *The Journal of Japanese Studies*, vol. 2, no. 1 (autumn 1975), pp. 1–28, esp. p. 20.

11. It should be pointed out that reforms "reflected the full range of the then American political spectrum: left, liberal, New Deal, mainline American, conservative, military." Herbert Passin, "The Occupation: Some Reflections," *Daedalus*, vol. 119, no. 3 (Summer 1990), pp. 107–129, esp. p. 114.

12. Robert Gilpin, "Introduction: Natural Scientists in Policymaking," in Robert Gilpin and Christopher Wright (eds), *Scientists and National Policy-Making* (New York: Columbia University Press, 1964), pp. 1–18, esp. pp. 4–5. Also see James L. Penick, Jr., Carroll W. Pursell, Jr., Morgan B. Sherwood, and Donald C. Swain (eds), *The Politics of American Science, 1939 to the Present* (Cambridge, MA: MIT Press, 1972 ed.), p. 120.

13. Supreme Commander for the Allied Powers (SCAP), General Headquarters, *History of the Nonmilitary Activities of the Occupation of Japan 1945–1951, Volume 54: Reorganization of Science and Technology in Japan* [hereafter referred to as *SCAP History*], p. 8, National Diet Library, Tokyo.

14. Yuasa, *History of Science*, pp. 290–91.

15. *SCAP History*, pp. 1–2.

16. Occupation Instructions no. 2, APO 500, September 25, 1945, Occupation Instructions, GHQ/SCAP, vol. 1, nos 1–4.

17. Yuasa, *History of Science*, p. 291.

18. *SCAP History*, p. 9.

19. APO 500, March 28, 1946, SCAPIN 830-A.

20. *SCAP History*, pp. 20–21.

21. *SCAP History*, p. 15.

22. *SCAP History*, p. 2.

23. Samuel K. Coleman, "Riken from 1945 to 1948: The Reorganization of Japan's Physical and Chemical Research Institute under the American Occupation," *Technology and Culture*, vol. 31, no. 2 (April 1990), pp. 228–50, esp. pp. 232–33.

24. Shūji Fukui has commented on the destruction of what was mistaken to be a "baby cyclotron" at Osaka, making the number of cyclotrons there two, and the total number of cyclotrons seized five. Discussed at Fermilab Workshop on the History of Japanese Particle Physics, May 6–7, 1985. The bulk of the proceedings of the workshop can be found in Laurie M. Brown, Rokuo Kawabe, Michiji Konuma, and Zirō Maki (eds), *Elementary Particle Theory in Japan, 1935–1960: Japan–USA Collaboration, Second Phase* (Kyoto: Yukawa Hall Archival Library, Research Institute for Fundamental Physics, 1988).

25. The newspaper also mistakenly reported that there were five cyclotrons. See Lindesay Parrott, "Five Cyclotrons Wrecked in Japan," *The New York Times*, November 24, 1945, p. 3.

26. William Lawren, *The General and the Bomb: A Biography of General Leslie R. Groves, Director of the Manhattan Project* (New York: Dodd, Mead and Co., 1988), p. 268.

27. Coleman, "Riken from 1945 to 1948," pp. 232–33.

28. Braw, *The Atomic Bomb Suppressed*, p. 111.

29. Letter, August 16, 1948, Yoshio Nishina to Ernest O. Lawrence, Ernest Orlando Lawrence Correspondence and Papers, BANC MSS 72/117C, Carton 9, Folder 38, The Bancroft Library, University of California, Berkeley.

30. Harry C. Kelly, Notes on Conference, December 26, 1946, GHQ/SCAP records, ESS (B) 11594, National Diet Library, Tokyo.

31. Yoshio Nishina, "Nihon saiken to kagaku" ("Science and the Reconstruction of Japan"), *Shizen*, no. 1 (May 1946), essay dated March 12, 1946, reprinted in *Shizen*, no. 300 (March 1971), pp. 15–18, esp. p. 17, my translation.

32. Yoshio Nishina, "Kokumin no jinkaku kōjō to kagaku gijutsu" ("Science, Technology and the Improvement of the Character of the People," dated October 20, 1946. Published in *Shizen* (December 1946) and reprinted in *Shizen*, no. 300 (March 1971), p. 19, my translation).

33. Scientific and Technical Section [sic Division] Chief to ESS Chief, "Proposal for Control of Japanese Scientific and Technological Acitivities," draft, May 2, 1946, GHQ/SCAP records, ESS (B) 11597, National Diet Library, Tokyo.

34. Science Council of Japan, *Annual Report, 1951–1958* (Tokyo, 1959), pp. 201–202.

35. SCAPIN 3452-A, March 20, 1947, SCAPIN-A, vol. 15, March–April 1947, National Diet Library, Tokyo. SCAPINs were SCAP instructions to the Japanese Government, apparently issued by ESS between September 4, 1945 and March 8, 1952. See Cohen, *Remaking Japan*, p. 475.

36. *SCAP History*, p. 60.
37. *SCAP History*, pp. 75–77.
38. Report, January 12, 1949, H.C. Kelly to C/S via General Staff Sections G-3, G-4, G-2,GHQ/SCAP, "Base Laboratories in Japan," check sheet, top secret, GHQ/SCAP records, TS 211, National Diet Library, Tokyo.
39. Robert Serber, *The Los Alamos Primer: The First Lectures on How to Build an Atomic Bomb* (Berkeley: University of California Press, 1992), p. 93.
40. See Shigeru Nakayama, "Kenkyū kichika jiken: Senryōka Nihon 'Ishidoru Rabi' " ("The Research Base Incident: Occupied Japan and Isidor Rabi"), *Kagaku Asahi* (September 1988), pp. 83–87, esp. p. 83.
41. Letter, December 8, 1948, I.I. Rabi to H.C. Kelly, GHQ/SCAP records, TS 211, National Diet Library, Tokyo.
42. Letter, December 8, 1948, Rabi to Kelly.
43. Report, September 3, 1948, H.C. Kelly to W.F. Marquat, "Utilization of Japanese Scientists by the United States," GHQ/SCAP records, TS 211, National Diet Library, Tokyo.
44. Report, Kelly to Marquat.
45. Contrary to Rabi's account, it appears that although American personnel were trained there and given access to facilities, the laboratory remained under the control of the British.
46. Report, December 10, 1948, I.I. Rabi to W.F. Marquat, "The Use of Japanese Research Facilities as an Advanced Base in the Event of Acute Emergency in the Far East," stamped top secret, GHQ/SCAP records, TS 210, National Diet Library, Tokyo.
47. "Leading Research Laboratories in Japan, Section I," top secret, GHQ/SCAP records, TS 210, National Diet Library, Tokyo.
48. "Leading Research Laboratories in Japan, Section II," pp. 15–16, top secret, GHQ/SCAP records, TS 211, National Diet Library, Tokyo.
49. Report, February 8, 1949, Col. Rash, General Staff Section G-4 to G-2, GHQ/SCAP, "Base Laboratories in Japan," check sheet, top secret, GHQ/SCAP records, TS 211, National Diet Library, Tokyo.
50. The name is only typed on the document. There may have been a simple typographical error. For background on Pash, see Christopher Simpson, *Blowback: America's Recruitment of Nazis and Its Effect on the Cold War* (London: Weidenfeld and Nicolson, 1988), pp. 152–53.
51. Richard Rhodes, *The Making of the Atomic Bomb* (London: Penguin Books, 1988), p. 606; Gregg Herken, *Brotherhood of the Bomb: The Tangled Lives and Loyalties of Robert Oppenheimer, Ernest Lawrence, and Edward Teller* (New York: Owl Books, Henry Holt and Co., 2003), p. 96.
52. Robert Jungk, trans. James Cleugh, *Brighter than a Thousand Suns: A Personal History of the Atomic Scientists* (New York: Harcourt, Brace and Co., 1958); David C. Cassidy, *Uncertainty: The Life and Science of Werner Heisenberg* (New York: W.H. Freeman and Co., 1992); Samuel A. Goudsmit, with an Introduction by David Cassidy, *Alsos*

(Woodbury, NY: AIP Press, 1996). Originally published New York: Henry Schuman, 1947; Mark Walker, *German National Socialism and the Quest for Nuclear Power, 1939–1949* (Cambridge: Cambridge University Press, 1989), pp. 153–54, 157–58.

53. Letter, ESS, SCAP to Deptartment of Army, March 19, 1949, draft, top secret, GHQ/SCAP records, TS 211, National Diet Library, Tokyo.

54. *SCAP History*, p. 77.

55. Karl T. Compton, "Mission to Tokyo," unpublished, typewritten manuscript, October 8, 1945, Institute Archives, MIT [T-N C738], courtesy of Prof R.W. Home, University of Melbourne. Also see James Phinney Baxter III, *Scientists Against Time* (Cambridge, MA: MIT Press, 1968), pp. 126, 410, 416.

56. L.N. Ridenour, "Science and the Federal Government," *Bulletin of the Atomic Scientists*, vol. 7 (February 1951), pp. 35–37, cited in Richard G. Hewlett and Francis Duncan, *A History of the United States Atomic Energy Commission, Vol. II, 1947–1952: Atomic Shield* (US Atomic Energy Commission, 1972).

57. For one account of the period leading up to the establishment of the Science Council, see Shigeru Nakayama, "The American Occupation and the Science Council of Japan," in E. Mendelsohn (ed.), *Transformation and Tradition in the Sciences: Essays in Honor of I. Bernard Cohen* (Cambridge: Cambridge University Press, 1984), pp. 353–69.

58. Minutes of Research Institutions Committee Meeting, March 4, 1947, Natural Resources Section, p. 2, GHQ/SCAP records, ESS (B) 11600, National Diet Library, Tokyo.

59. "Progress Report of Chairman, Renewal Committee for the Organization of Science at 1st General Meeting of the Science Council of Japan," January 20, 1949, GHQ/SCAP records, ESS (B) 11631, National Diet Library, Tokyo.

60. "The Progress of the Renewal Movement Since the End of the War." Typewritten report on National Research Council stationery, August 25, 1947. GHQ/SCAP records, ESS (B) 11599, National Diet Library, Tokyo. See also Japan Academy, *Nihon Gakushiin shōshi (A Short History of the Japan Academy)* (Tokyo: Japan Academy, 1980).

61. "Address to Representatives of Japanese Scientific Organization by Brig. John W. O'Brien, chief, Scientific and Technical Division, ESS, GHQ, SCAP, Central Liaison Office, June 5, 1946, 1:30 pm," in Japanese Government, Foreign Office, Division of Special Records, "Documents Concerning the Allied Occupation and Control of Japan," vol. 5, I, 1.2.0.5 (November 1945–March 1949), pp. 15–16, Waseda University Microfilm Center.

62. Draft of "Summary of Activities of Scientific and Technical Division in Science Reorganization," GHQ/SCAP records, ESS (B) 11598. National Diet Library, Tokyo.

63. *SCAP History*, pp. 31–33.
64. "JASL in the Past, Present and in Future," JASL report, May 29, 1947, p. 1, GHQ/SCAP records, ESS (B) 11600, National Diet Library, Tokyo.
65. *SCAP History*, p. 48.
66. "Classified Topics Expected for the Discussions." Undated memorandum on National Research Council stationery, p. 1, GHQ/SCAP records, ESS (B) 11599, National Diet Library, Tokyo.
67. Minutes of Research Institutions Committee Meeting, Natural Resources Section, March 4, 1947, p. 2, GHQ/SCAP records, ESS (B) 11600, National Diet Library, Tokyo.
68. "Summary of Activities Connected with Japanese Science Reorganization" prepared by the Scientific and Technical Division, ESS, for the assistance of the Scientific Advisory Group, July 1947, pp. 9–10. GHQ/SCAP records, ESS (B) 11609, National Diet Library, Tokyo.
69. Minutes of Meeting on Scientific Reorganization, April 2, 1947, Scientific and Technical Division, Economic and Scientific Section, p. 3, GHQ/SCAP records, ESS (B) 11600, National Diet Library, Tokyo.
70. Satoshi Watanabe, "Suggestions for Rehabilitation of Scientific Activities in Japan," July 1947, Tokyo, p. 2, GHQ/SCAP records, ESS (B) 11604, National Diet Library, Tokyo.
71. *SCAP History*, p. 51.
72. *SCAP History*, p. 51.
73. *SCAP History*, pp. 57, 61.
74. Nakayama, "The American Occupation."
75. Minutes of the Eighth General Meeting of the Renewal Committee of Scientific Systems, GHQ/SCAP records, ESS (B) 11590, National Diet Library, Tokyo.
76. Letter, May 13, 1948, Nagaoka, President, Japan Academy to O'Brien, ESS, GHQ/SCAP records, ESS (B) 11604, National Diet Library, Tokyo.
77. See "Man and Man (36): Kaneshige and Yamada," *Osaka Jiji Shinpō*, September 10, 1948, English translation, GHQ/SCAP records, ESS (B) 11626, National Diet Library, Tokyo.
78. President Harry S. Truman, Congressional Record, Appendix (August 15, 1947), A4442–A4443, reprinted in James L. Penick, Jr. et al. (eds), *The Politics of American Science, 1939 to the Present* (Cambridge, MA: MIT Press, 1972 ed.), pp. 134–37, esp. p. 135.
79. Law 121, 1948, "The Science Council of Japan Law." GHQ/SCAP records, ESS (B) 11609; "Progress Report of Chairman, Renewal Committee for the Organization of Science."
80. For details of the voting, see Minutes of First General Meeting of JSC, January 20, 1949 at the Japan Academy Auditorium, GHQ/SCAP records, ESS (B) 11634.
81. *SCAP History*, p. 72.

82. Letter, January 12, 1949, SCAP to Dept of Army, with memorandum by H.C. Kelly, ESS (B) 11608; J.W. O'Brien, memorandum on the Scientific and Technical Administrative Committee of the Japan Science Council, May 3, 1948, GHQ/SCAP records, ESS (B) 11605, National Diet Library, Tokyo.

83. "Report on Missions and Accomplishments of Scientific and Technical Division, Economic and Scientific Section," September 16, 1949, GHQ/SCAP records, ESS (A)-09796, National Diet Library, Tokyo.

84. Samuel K. Coleman, "Riken from 1945 to 1948: The Reorganization of Japan's Physical and Chemical Research Institute under the American Occupation," *Technology and Culture*, vol. 31, no. 2 (April 1990), pp. 228–50, esp. pp. 238–39, 244. See also R.W. Home and Masao Watanabe, "Forming New Physics Communities: Australia and Japan, 1914–1950," *Annals of Science*, vol. 47 (1990), pp. 317–45, esp. p. 334.

85. Yoshio Nishina, "Kabushiki Gaisha Kagaku Kenkyūjo no shimei" ("The Mission of the Scientific Research Institute"), April 7, 1948, *Shizen* (June 1948), reprinted in *Shizen*, no. 300 (March 1971), p. 21.

86. Coleman, "Riken from 1945 to 1948," pp. 228–50. This institution continues under the same name, the Institute for Physical and Chemical Research, but is now almost totally supported by the Japanese government as a nonprofit research corporation of the Science and Technology Agency. This is in contrast to its prewar and wartime role as an institute that combined basic research with industrial applications, and which derived income from their commercialization in the form of manufactured goods, such as machine tools, electrical equipment, optical instruments, chemicals, and pharmaceuticals.

87. For one account, see Hidehiko Tamaki, "Kagaku Kenkyūjo to Nishina sensei" ("The Scientific Research Institute and Dr. Nishina"), *Kagaku*, vol. 21, no. 4 (April 1951), pp. 212–15; also Coleman, "Riken from 1945 to 1948," pp. 244–45.

88. Letter, August 17, 1948, Y. Nishina to N. Bohr at the Institute for Advanced Study, Princeton University, in *Y. Nishina's Correspondence with N. Bohr and Copenhageners, 1928–1949* (Tokyo: Nishina Memorial Foundation, 1984), pp. 62–63.

89. Letter, January 3, 1947, Y. Nishina to E.O. Lawrence, Ernest Orlando Lawrence Correspondence and Papers, BANC MSS 72/117C, Carton 9, Folder 38, The Bancroft Library, University of California, Berkeley.

90. Robert Oppenheimer, writing in *Technology Review* (ca. February 1948), reported in *Time* (February 23, 1948), clipping, ESS files, ESS (E) 06380, GHQ/SCAP records, National Diet Library, Tokyo.

91. Letter, ca. late 1945, Ryōkichi Sagane to E.O. Lawrence, in Japanese, accompanied by translation and letter, January 4, 1946, Maj. Gen. Leslie R. Groves to E.O. Lawrence, Ernest Orlando Lawrence

Correspondence and Papers, BANC MSS 72/117C, Carton 8, Folder 7, The Bancroft Library, University of California, Berkeley.

92. Letter, July 15, 1946, Nishina to Lawrence, Ernest Orlando Lawrence Correspondence and Papers, BANC MSS 72/117C, Carton 9, Folder 38, The Bancroft Library, University of California, Berkeley.

93. Letter, November 22, 1946, H.C. Kelly, Deputy Chief, Scientific and Technical Division, ESS, GHQ/SCAP, to E.O. Lawrence, Ernest Orlando Lawrence Correspondence and Papers, BANC MSS 72/117C, Carton 9, Folder 39, The Bancroft Library, University of California, Berkeley.

94. Letters, August 19, August 26, and October 8, 1949, Nishina to Bohr, in *Y. Nishina's Correspondence with N. Bohr and Copenhageners*, pp. 65–66, 66–67, and 67–70.

95. Letter, September 25, 1950, Nishina to Lawrence, Ernest Orlando Lawrence Correspondence and Papers, BANC MSS 72/117C, Carton 9, Folder 38, The Bancroft Library, University of California, Berkeley.

96. Takatoshi Ito, *The Japanese Economy* (Cambridge, MA: MIT Press, 1992), p. 68.

97. Letter, September 25, 1950, Nishina to Lawrence.

98. Braw, *The Atomic Bomb Suppressed*, p. 81. The "Red Purge" in the private sector involved 11,000 persons being made redundant during July–October 1950. Cohen, *Remaking Japan*, p. 449. The purge in the communications area is said to have occurred much earlier, between August 1947 and May 1948. See Jay Rubin, "From Wholesomeness to Decadence: The Censorship of Literature under the Allied Occupation," *Journal of Japanese Studies*, vol. 11, no. 1 (1985), pp. 71–103, esp. p. 95.

99. Translation of excerpt from article by Yoshio Nishina, published in the magazine *Minron*, ca. 1948, GHQ/SCAP records, ESS (E)-06392, pp. 5–6.

100. Yoshio Nishina, "Genshiryoku to heiwa" ("Atomic Energy and Peace"), in Yomiuri Shinbun, Science Section (ed.), *Genshiryoku bunmei (Atomic Culture)* (Tokyo: Takayama Shoin, 1949), pp. 1–8, esp. pp. 1–2, my translation.

101. Fumio Yamazaki, "A Short Biography of Dr. Yoshio Nishina," in Sin-itirō Tomonaga and Hidehiko Tamaki (eds), *Nishina Yoshio: Denki to kaisō (Yoshio Nishina: Biography and Reminiscences)* (Tokyo: Misuzu Shobō, 1952), pp. 3–15, esp. p. 14.

102. Satoshi Aoki, *Dokyumento: Nakasone Famarii (A Documented Account: The Nakasone "Family")* (Tokyo: Akebi Shobō, 1986), pp. 33, 42.

103. Nathaniel B. Thayer, *How the Conservatives Rule Japan* (Princeton: Princeton University Press, 1969, 1973), p. 7.

104. Citation from *Gendai* (February 1978), in Michitaka Kido, *Nakasone no kokka kaizō keikaku hihan (Criticism of Nakasone's*

Plans for Rebuilding the Nation) (Tokyo: Senshinsha, 1986), pp. 129–30.

105. Satoshi Aoki, *A Documented Account*, p. 43.

106. Kenzō Uchida, "Nakasone: Rear Guard or Vanguard?," *Japan Echo*, vol. 10, no. 1 (1983), pp. 19–23, esp. p. 22.

107. Michitaka Kido, *Nakasone no kokka kaizō keikaku hihan* (*Criticism of Nakasone's Plans for Rebuildinq the Nation*) (Tokyo: Senshinsha, 1986), p. 130.

108. Takeo Ōta, secretary to Nakasone, cited in Thayer, *How the Conservatives Rule Japan*, p. 90.

109. John Welfield, *An Empire in Eclipse: Japan in the Postwar American Alliance System: A Study in the Interaction of Domestic Politics and Foreign Policy* (London: Athlone Press, 1988), pp. 26–27.

110. Michael Schaller, *The American Occupation of Japan: The Origins of the Cold War in Asia* (New York: Oxford University Press, 1985), pp. 9–19, 92–94.

111. Welfield, *An Empire in Eclipse*, pp. 66–67.

112. Gerald L. Curtis, *The Japanese Way of Politics* (New York: Columbia University Press, 1988), pp. 10–11.

113. Takeshi Ishida and Ellis S. Krauss, "Democracy in Japan: Issues and Questions," in Takeshi Ishida and Ellis S. Krauss (eds), *Democracy in Japan* (Pittsburgh: University of Pittsburgh Press, 1989), pp. 3–16, esp. p. 10.

114. Uchida, "Nakasone: Rear Guard or Vanguard?," p. 21.

115. Uchida, "Nakasone: Rear Guard or Vanguard?," p. 21.

116. Ken'ichi Nakamura, "Militarization of Post-war Japan," in Yoshikazu Sakamoto (ed.), *Asia: Militarization and Regional Conflict* (Tokyo: United Nations University; London: Zed Books, 1988), pp. 81–100, esp. p. 99.

117. Masao Kobayashi, *Saishō Nakasone Yasuhiro: Naikaku sōri daijin e no sokuseki* (*Prime Minister Yasuhiro Nakasone: The Road to the Prime Ministership*) (Tsu: Ise Shinbunsha, 1985), pp. 703–709.

118. Kenneth B. Pyle, "In Pursuit of a Grand Design: Nakasone Betwixt the Past and the Future," *The Journal of Japanese studies*, vol. 13, no. 2 (Summer 1987), pp. 243–70, esp. p. 251; Nakamura, "Militarization of Post-war Japan," pp. 81–100, esp. p. 99.

119. Kobayashi, *Prime Minister Yasuhiro Nakasone*, pp. 703–709, esp. p. 703.

120. Michael Schaller, *Douglas MacArthur: The Far Eastern General* (New York: Oxford University Press, 1989, 1990).

121. Harries and Harries, *Sheathing the Sword*, pp. 235–36.

122. Welfield, *An Empire in Eclipse*, p. 72.

123. Yasuhiro Nakasone, "Kagaku Gijutsu Chō setsuritsu made no omoide" ("Memories up until the Establishment of the Science and Technology Agency"), in Science and Technology Agency, *Kagaku Gijutsu Chō: 30 nen no ayumi* (*The Science and Technology Agency: A 30 Year History*) (Tokyo: Sōzō, 1986), pp. 94–95.

124. Harries and Harries, *Sheathing the Sword*, p. 236.
125. Curtis, *The Japanese Way of Politics*, p. 13.

CHAPTER 4 PHYSICISTS ON THE LEFT: SAKATA AND TAKETANI

1. See this discussed in the context of scientism in Hitoshi Yoshioka, *Kagakusha wa kawaru ka: Kagaku to shakai no shisōshi (Will Scientists Change?: An Intellectual History of Science and Society)* (Tokyo: Shakai Shisōsha, 1984), esp. p. 65.
2. Nobuko Sakata, "Sobo Koganei Kimiko no tegami (2)" ("The Letters of My Grandmother, Kimiko Koganei, Part 2"), *Chijiku (Earth's Axis)*, no. 14 (March 8, 1995), pp. 76–87.
3. Shōichi Sakata, *Kagakusha to shakai, Ronshū 2 (Scientists and Society, Collected Papers, Vol. 2)* (Tokyo: Iwanami Shoten, 1972), pp. 422–23; Shunkichi Hirokawa and Shuzō Ogawa, "Shōichi Sakata: His Physics and Methodology," *Historia Scientiarum*, no. 36 (1989), pp. 67–81, esp. p. 67.
4. Sakata, *Scientists and Society*, p. 423; Hirokawa and Ogawa, "Shōichi Sakata," pp. 68–69.
5. Sakata, *Scientists and Society*, p. 424.
6. Sakata, *Scientists and Society*, pp. 424–25.
7. For more information on Kakiuchi, see Tatsumasa Dōke, "Establishment of Biochemistry in Japan," *Japanese Studies in the History of Science*, no. 8 (1969), pp. 145–53 and Umeo Mogami and Kuro Iseki (eds), *Who's Who in "Hakushi" in Great Japan* (Tokyo: Hattensha, 1926), vol. 3, "Igaku Hakushi" (part 2), p. 364.
8. For details of Koganei, see Yasuo Yokoo, "Koganei Yoshikiyo: Nihon kaibōgaku no kusawake" ("Yoshikiyo Koganei: Pioneer of Anatomy in Japan"), in *Echigo no unda Nihonteki jinbutsu (Famous Japanese Born in Echigo)*, pp. 96–112, Nagaoka City Library collection, Japan. Also, K.R. Iseki (ed.), *Who's Who in "Hakushi" in Great Japan, 1888–1922*, vol. 2, "Igaku Hakushi" (part 1), pp. 8–9.
9. Chūbu Nihon Shinbunsha (ed.), *Gendai no keifu: Nihon o ugokasu hitobito (Contemporary Lineages: The "Movers" of Japan)* (Tokyo: Tokyo Chūnichi Shinbun Shuppankyoku, 1965), pp. 296–97. For the Mori family tree, see Takashi Hayakawa, *Nihon no jōryū shakai to keibatsu (Japan's Social Elite and Matrimonial Influence)* (Tokyo: Kadokawa Shoten, 1983), p. 240. For other biographical information, see Iseki, *Who's Who*, pp. 21–22.
10. For a discussion on the incidence of academics marrying daughters of professors and how family connections facilitated academic careers, see James R. Bartholomew, *The Formation of Science in Japan: Building a Research Tradition* (New Haven: Yale University Press, 1989), pp. 168–76.
11. Hirokawa and Ogawa, "Shōichi Sakata," pp. 69–73. Sakata, *Scientists and Society*, pp. 425–27.

12. Shōichi Sakata, *Shōichi Sakata: Scientific Works* (Nagoya: Publication Committee of Scientific Papers of Prof. Shōichi Sakata, 1977), p. ix.
13. J.D. Bernal, *The Social Function of Science* (London: Routledge, 1939), pp. 415–16.
14. Bernal, *The Social Function*, p. 267.
15. Mituo Taketani, *Shisō o oru* (*The Interweaving of Ideas*) (Tokyo: Asahi Shinbunsha, 1985), pp. 77–86.
16. Taketani, *The Interweaving of Ideas*, pp. 3–4, 6, 12–13, 28–29.
17. Taketani, *The Interweaving of Ideas*, pp. 42, 48, 56.
18. Acquaintances, which Taketani made at this time of unrest, would later be renewed through the magazine *Sekai bunka* (*World Culture*).
19. Taketani, *The Interweaving of Ideas*, pp. 57–58.
20. Taketani, *The Interweaving of Ideas*, pp. 63, 69–75.
21. For details of the activities of the Special Higher Police, see Elise K. Tipton, *The Japanese Police State: The Tokko in Interwar Japan* (North Sydney: Allen and Unwin, 1990).
22. Taketani, *The Interweaving of Ideas*, pp. 76–84.
23. Taketani, *The Interweaving of Ideas*, pp. 86–91.
24. In this process, the uranium isotopes were separated by thermal diffusion of uranium hexafluoride, the only gaseous compound of uranium. A thermal column consisting of 17-feet concentric pipes, with the inner pipe electrically heated to 750°F and the outer pipe water-cooled, was built to separate the two. For one account, see Kei-ichi Tsuneishi, "Riken ni okeru uran bunri no kokoromi" ("The Attempt to Separate Uranium at Riken"), *Nihon Butsuri Gakkaishi* (*Bulletin of the Physicial Society of Japan*), vol. 45, no. 11 (November 1990), pp. 820–25. Also see Richard Rhodes, *The Making of the Atomic Bomb* (London: Penguin Books, 1988), pp. 580–82; Robert K. Wilcox, *Japan's Secret War* (New York: William Morrow, 1985), pp. 94–95.
25. Taketani, *The Interweaving of Ideas*, pp. 92–97, 104–106.
26. Taketani, *The Interweaving of Ideas*, pp. 107–108.
27. Taketani, *The Interweaving of Ideas*, pp. 118–19.
28. See Alice Kimball Smith, *A Peril and a Hope: The Scientists' Movement in America, 1945–47* (Chicago: University of Chicago Press, 1965); William McGucken, *Scientists, Society and State: The Social Relations of Science Movement in Great Britain, 1931–1947* (Colombus: Ohio State University Press, 1984).
29. Masanori Ōnuma, Yōichirō Fujii, and Kunioki Katō, *Sengo Nihon kagakusha undōshi, jyō* (*The History of the Postwar Japanese Scientist Movement, Part One*) (Tokyo: Aoki Shoten, 1975), pp. 23–24.
30. Ōnuma et al., *The History*, p. 25.
31. A list of foundation members and office bearers is included in Hideomi Tuge, *Minka to watakushi: Sengo hito kagakusha no ayumi* (*The Association of Democratic Scientists and I: The Postwar Story of a Scientist*) (Tokyo: Keisō Shobō, 1980), p. 294.

32. There are strong parallels with England. See Gary Werskey, *The Visible College* (London: Allen Lane, 1978).

33. Confidential intercept of Kōichirō Ichikawa (c/Minka) to Kansei Inoue (Imperial Petroleum), June 15, 1949, Civil Censorship Detachment, CIS 03977, GHQ/SCAP records, National Diet Library, Tokyo.

34. Jay Rubin, "From Wholesomeness to Decadence: The Censorship of Literature under the Allied Occupation," *Journal of Japanese Studies*, vol. 11, no. 1 (1985), pp. 71–103, esp. p. 72.

35. Rubin, "From Wholesomeness to Decadence," pp. 74–75.

36. Bruno Latour, *Pandora's Hope: Essays on the Reality of Science Studies* (Cambridge, MA: Harvard University Press, 1999), p. 87.

37. For an explanation of Taketani's concept of technology, see Yoshirō Hoshino, "On Concepts of Technology," in Shigeru Nakayama, David L. Swain, and Eri Yagi (eds), *Science and Society in Modern Japan* (Tokyo: University of Tokyo Press, 1974), pp. 39–50.

38. Taketani, *The Interweaving of Ideas*, pp. 120–21.

39. Taketani, *The Interweaving of Ideas*, pp. 128–29.

40. Translation by Allied Translator and Interpreter Service, Military Intelligence Section, GHQ of Department of Science and Technology of the Japan Communist Party, "Shortcomings of Science and Technology in Japan and the Duty of the Communists," *Zen-ei* (*Vanguard*) (November 1946), GHQ/SCAP records, ESS (E) 06393, National Diet Library, Tokyo.

41. Department of Science and Technology.

42. "Gakumon to shisō no jiyū no tame ni" (1950), in Sakata, *Scientists and Society*, p. 73.

43. Department of Science and Technology, "Shortcomings of Science and Technology in Japan."

44. Taketani, *The Interweaving of Ideas*, pp. 122, 126–27.

45. Mituo Taketani, "Democratic Revolution of Japan and Technologists," *Gijutsu* (*Technology*), vol. 5, no. 2 (March 1946), pp. 3–5, abridged translation, GHQ/SCAP records, ESS (E) 06393, National Diet Library, Tokyo.

46. Taketani, *The Interweaving of Ideas*, pp. 122, 126–27.

47. Mitutomo Yuasa, *Kagakushi* (*The History of Science*) (Tokyo: Tōyō Keizai Shinpōsha, 1961), p. 292.

48. The Physical Society of Japan was established from part of what had previously been the Physico-Mathematical Society of Japan. A particle theory sectional meeting was formed at the same time. This group later divided up into a particle theory and a nuclear theory group. Tetu Hirosige, *Sengo Nihon no kagaku undō* (*Postwar Japan Science Movement*) (Tokyo: Chūō Kōronsha, 1960), pp. 181–82. Bi-annual meetings of the Elementary Particle Theory Forum (as part of Physical Society conferences) have provided and do provide occasions for group decision-making, while administrative work has been shared

by an "Elementary Particle Theory Group Secretariat," which rotates every 3–4 months from campus to campus. In addition to such meetings, the group also assists in the organization of research meetings at RIFP and INS.

49. Bernal, *The Social Function of Science*, p. xiii.
50. Bernal, *The Social Function of Science*, p. 265.
51. Sakata, *Scientists and Society*, p. 428.
52. Ōnuma et al., *Postwar Scientist Movement, Part One*, pp. 35, 37–38.
53. J.D. Bernal, *The Freedom of Necessity* (London, 1949), pp. 311–12, cited in Werskey, *The Visible College*, p. 274.
54. Draft translation by "George" of Shōichi Sakata, "Abolish the Sectionalism: On the Democratization of Research System," *Tokyo Imperial University Newspaper*, no. 1004 (Wednesday, November 13, 1946), GHQ/SCAP records, ESS (E) 06393, National Diet Library, Tokyo.
55. Letter, September 19, 1945, Harold Fair, Assistant Adjutant General, SCAP, to Imperial Japanese Government regarding press code for Japan, in Eizaburō Okuizumi (ed.), *User's Guide to the Gordon W. Prange Collection, East Asia Collection, McKeldin Library, University of Maryland at College Park, Part 1: Microfilm Edition of Censored Periodicals, 1945–1949* (Tokyo: Yushodō Booksellers Ltd., 1982), p. 30.
56. Okuizumi, *User's Guide*.
57. See, e.g., "Information slips," September 1949, Civil Censorship Detachment (Press, Pictorial, Broadcast Division) on the Democratic Scientists' Association, GHQ/SCAP records, CIS-03978, National Diet Library, Tokyo.
58. Ichikawa to Inoue, June 15, 1949.
59. Okuizumi, *User's Guide*, pp. 8–12.
60. Kyodo-U.P., "Red Controlled Japan Held Threat to Far East," *Nippon Times*, June 25, 1948, ESS file no. 0603.1, ESS (D)-03157, GHQ/SCAP records, National Diet Library, Tokyo.
61. "Anti-Communist Drive By Government Denied: Equal Treatment Under Law will be Accorded Reds," *Nippon Times*, June 10, 1948, GHQ/SCAP records, ESS (D) 03157, National Diet Library, Tokyo.
62. Civil Censorship Detachment, Intercept 11525 8/31 of Letter, August 18, 1948, Sakata to Nippon Butsuri Gakkai (Physical Society of Japan), in Japanese, translated August 20, 1948, GHQ/SCAP records, ESS (A) 09912, National Diet Library, Tokyo.
63. Translation of statement by Sakata, ca. June 1949, GHQ/SCAP records, ESS (E) 06366, National Diet Library, Tokyo.
64. Kyodo-A.P., "President Encourages U.S. Capital to Do Part In Battle Against Reds," *Nippon Times*, January 22, 1949, ESS file no. 0603.1, ESS (D) 03157, GHQ/SCAP records, National Diet Library, Tokyo.
65. Bernal, *The Social Function of Science*, p. 415.
66. For example, see Shōichi Sakata, "Taketani's Three-Stage Theory," *Supplement of the Progress of Theoretical Physics*, no. 50 (1971),

pp. 9–11; originally published as part of Shōichi Sakata, "Course of the Development of the Yukawa Theory," in Japanese, written in 1948, *Shizen* (*Nature*) (1949), and also included in Shōichi Sakata, *Butsurigaku to hōhō* (*Physics and Method*) (Tokyo: Iwanami Shoten, 1951).

67. Sakata, *Scientists and Society*, p. 48.
68. The survey was conducted in 1951. See "Hiroi kanten ni tatte: Iwayuru 'Genshiryoku kenkyū mondai' o megutte" ("Looking at the So-called Atomic Energy Research Problem From a Wide Point of View") (1953); Sakata, *Scientists and Society*, pp. 126–30, esp. pp. 128–29.
69. Sakata, *Scientists and Society*, p. 421.
70. Sakata, *Scientists and Society*, pp. 428–29.
71. Translation of summary of interview with Mituo Taketani, *Minshū hyōron* (*People's Review*), June 1949, GHQ/SCAP records, Item no. 10, ESS (E) 06394, National Diet Library, Tokyo.
72. This is discussed in Paul Forman, "Social Niche and Self-Image of the American Physicist," in Michelangelo De Maria, Mario Grilli, and Fabio Sebastiani (eds), *The Restructuring of Physical Sciences in Europe and the United States, 1945–1960* (Singapore: World Scientific, 1989), pp. 96–104, esp. p. 98.
73. See Monica Braw, trans. Seiitsu Tachibana, *Ken-etsu: Kinjirareta genbaku hōdō* (*American Censorship in Japan, 1945–1949: The Atomic Bomb Suppressed*) (Tokyo: Jiji Tsūshinsha, 1988).
74. "The Smyth Report" is the official account of the history of the development of the American atomic bomb, written by the experimental physicist Henry DeWolf Smyth, distributed to the press on August 11, 1945 and later sold in book form to the public. It has recently been reissued as: Henry DeWolf Smyth, *Atomic Energy for Military Purposes: The Offical Report on the Development of the Atomic Bomb Under the Auspices of the United States Government, 1940–1945* (Stanford: Stanford University Press, 1989).
75. Taketani, *The Interweaving of Ideas*, pp. 132–34.
76. Japan Atomic Industrial Forum, *Nihon no genshiryoku: 15 nen no ayumi, jyō* (*Atomic Energy in Japan: A 15 Year History, Part 1*) (Tokyo: Japan Atomic Industrial Forum, 1971), pp. 12–14.
77. Taketani, *The Interweaving of Ideas*, p. 144.
78. Spencer R. Weart, *Scientists in Power* (Cambridge, MA: Harvard University Press, 1979), pp. 216, 219, 222, 242.
79. "Genshiryoku to torikumu" ("Grappling with Atomic Energy") (1952), in Sakata, *Scientists and Society*, pp. 116–22, esp. p. 121.
80. John Beatty, "Scientific Collaboration, Internationalism, and Diplomacy: The Case of the Atomic Bomb Casualty Commission," *Journal of the History of Biology*, vol. 26, no. 2 (Summer 1993), pp. 205–31, esp. p. 219.
81. Mituo Taketani, Yōichi Fujimoto, and Takeshi Kawai, " 'Kagaku jidai' o kangaeru" ("Reflections on the Scientific Age"), compilation of

three articles which appeared in *Ekonomisuto*, June 25, July 2, July 9, 1968. Reprinted in Mituo Taketani, *Gijutsu to kagaku gijutsu seisaku: Bunmei, daigaku mondai, dokusen shihon (Technology and Science Policy: Civilization, University Problems, and Monopolistic Capital) (The Collected Works of Mituo Taketani, Vol. 3)* (Tokyo: Keisō Shobō, 1976), pp. 122–62, esp. p. 142.

82. Mituo Taketani, "Himotsuki genshiryoku ni keikoku suru" ("Nuclear Power has Strings Attached: A Warning"), *Ekonomisuto*, June 10, 1955, special supplement, reprinted in Taketani, *Technology and Science Policy*, pp. 10–24, esp. p. 20.

83. Manabu Hattori, "Genshiryoku shisetsu no anzensei to Taketani sensei" ("Dr. Taketani and the Safety of Nuclear Power Facilities"), *Taketani Mituo chosakushū geppō 2 (The Collected Works of Mituo Taketani, Monthly Bulletin No. 2)* (Tokyo: Keisō Shobō, August 1968), p. 1.

84. Daniel Iwao Okimoto, "Ideas, Intellectuals, and Institutions: National Security and the Questions of Nuclear Armament in Japan," 2 vols., unpublished PhD dissertation, University of Michigan, 1978, vol. 1, pp. 23–24.

85. Okimoto, "Ideas, Intellectuals, and Institutions," pp. 23–24.

86. Sakata, *Scientists and Society*, p. 112.

87. Sakata's association with the JAEC did not last long for he resigned on November 10, 1959. Sakata, *Scientists and Society*, pp. 112, 429–30.

88. Supreme Commander for the Allied Powers (SCAP), General Headquarters, *History of the Nonmilitary Activities of the Occupation of Japan 1945–1951, Volume 54: Reorganization of Science and Technology in Japan*, National Diet Library, Tokyo, p. 97.

89. Science Council of Japan, *Annual Report, 1951–1958* (Tokyo, 1959), pp. 201–202.

90. Science Council of Japan, pp. 109–111, 228.

91. "Genshiryoku ni tsuite no uttae: Sutokuhorumu nite" ("An Appeal Regarding Atomic Energy: At Stockholm") (1956), in Sakata, *Scientists and Society*, pp. 242–45.

92. For details of the meeting, see Eugene Rabinowitch, "The Third Pugwash Conference," in Morton Grodzins and Eugene Rabinowitch (eds), *The Atomic Age: Scientists in National and World Affairs* (New York: Basic Books, 1963), pp. 552–57.

93. "Kagaku Gijutsu Kaigi no setchi to genshiryoku no anzensei o megutte" ("The Establishment of the Science and Technology Council and the Safety of Atomic Energy") (1958), in Sakata, *Scientists and Society*, pp. 150–59, esp. p. 151.

94. Sakata, *Scientists and Society*, p. 430.

95. "Statement of the Kyoto Conference of Scientists," May 9, 1963, signed by Bokuro Eguchi, Osamu Kuno, Yositaka Mimura, Yasuo Miyake, Mokichirō Nogami, Shōichi Sakata, Kiyosi Sakuma, Hiroshi

Suekawa, Shinjirō Tanaka, Sin-itirō Tomonaga, and Hideki Yukawa, (printed statement, translated from the Japanese original, Sakata Archival Library, Nagoya University).

96. For papers presented by the Japanese delegation, see *Papers Presented at the 1964 Peking Symposium, Vol. 1: Natural Science* (Peking: Scientific and Technical Association of the People's Republic of China, 1965).

97. Toshiyuki Toyoda, "Scientists Look at Peace and Security," *Bulletin of the Atomic Scientists*, vol. 40, no. 2 (February 1984), pp. 16–19, esp. p. 18.

98. Bertrand Russell to Sakata, on Bertrand Russell Peace Foundation letterhead, April 14, 1967; Russell to Sakata, on International War Crimes Tribunal letterhead, April 18, 1967, Sakata Archival Library, Nagoya University.

99. "Kagaku no ronri to seiji no ronri: Gakujutsu Kaigi nijūnen" ("The Logic of Science and the Logic of Government: Twenty Years of the Science Council of Japan") (1968), in Sakata, *Scientists and Society*, pp. 389–93.

100. Taketani et al., "Reflections on the Scientific Age," pp. 122–23.

101. Sakata, *Scientists and Society*, p. 430.

102. This was also the case in the United States. See Brian Balogh, *Chain Reaction: Expert Debate and Public Participation in American Commmercial Nuclear Power, 1945–1975* (Cambridge: Cambridge University Press, 1991).

103. Elementary Particle Theory Group, Nagoya University Branch, "A Short History of the Research Group of the Theory of Elementary Particles of Japan," mimeographed article, Sakata Archival Library, Nagoya University, p. 5.

104. Yasuhisa Katayama and Eiji Yamada, "Soryūshiron Guruupu no mondai" ("The Problems of the Elementary Particle Theory Group"), in Yoshirō Hoshino (ed.), *Sengo Nihon shisō taikei 9: Kagaku gijutsu no shisō (An Outline of Postwar Thought, Vol. 9: Scientific and Technological Thought)* (Tokyo: Chikuma Shobō, 1971), pp. 335–60, esp. p. 341.

105. For a list of these members, see Yoshinori Kaneseki, "The Elementary Particle Theory Group," in Nakayama et al., *Science and Society in Modern Japan*, pp. 221–52, esp. pp. 222–29. For an estimate of between 150 and 200 researchers, see Katayama and Yamada, "The Problems of the Elementary Particle Theory Group," p. 346.

106. Katayama and Yamada, "The problems of the Elementary Particle Theory Group," p. 345.

107. Elementary Particle Theory Group, "A Short History," p. 5.

108. Hirosige, *Postwar Japan Science Movement*, pp. 184–85.

109. Sakata, *Scientists and Society*, p. 84.

110. Shinobu Nagata, "Scientific and Social Activities of 'Soryūshiron' Group in Japan," abstract, *Contributions at the 1964 Peking Symposium*, Gen.:019 (Peking: Scientific and Technical Association of the People's Republic of China, 1965), pp. 681–83.

111. Elementary Particle Theory Group, "A Short History," p. 6.

112. Elementary Particle Theory Group, "A Short History," p. 7.

113. Katayama and Yamada, "The Problems of the Elementary Particle Theory Group," p. 347.

114. Elementary Particle Theory Group, "A Short History," p. 6. See also Katayama and Yamada, "The Problems of the Elementary Particle Theory Group," pp. 349–50.

115. Taketani, *The Interweaving of Ideas*, pp. 135–36.

116. Kiyonobu Itakura, "Kagakusha no jishuteki na soshiki" ("An Independent Organization of Scientists"), in Mituo Taketani (ed.), *Shizen kagaku gairon, daiikkan: Kagaku gijutsu to Nihon shakai (An Introduction to the Natural Sciences, Vol. 1: Science, Technology and Japanese Society)* (Tokyo: Keisō Shobō, 1962), pp. 155–73.

117. Hirosige, *Postwar Japan Science Movement*, p. 182.

118. Hirosige, *Postwar Japan Science Movement*, p. 300. Around 1958, membership exceeded 300 and, by 1960, reached a total of 360.

119. History of Science Society of Japan (ed.), *Nihon kagaku gijutsu shi taikei, 5: Tsūshi 5 (An Outline of the History of Japanese Science and Technology, Vol. 5: General History no. 5)* (Tokyo: Daiichi Hōki, 1964), p. 33.

120. Hirosige, *Postwar Japan Science Movement*, p. 180.

121. Michiji Konuma, "Social Aspects of Japanese Particle Physics in the 1950s," in Laurie M. Brown, Max Dresden, and Lillian Hoddeson (eds), *Pions to Quarks: Particle Physics in the 1950s* (Cambridge: Cambridge University Press, 1989), pp. 536–48, esp. p. 542.

122. For an account of the social role of the Elementary Particle Theory Group, see Japan Physics Committee for 1964 Peking Symposium, Kyoto Working Group, "Scientific and Social Activities of 'Soryūshiron' Group in Japan," in *Papers Presented at the 1964 Peking Symposium, Vol. 1*, pp. 291–302.

123. Elementary Particle Theory Group, "A Short History," p. 9.

124. Michiji Konuma, Chieko Masuzawa, and Yoshio Takada, "Resumption of International Relationship of Japanese Particle Physicists after World War II," *Historia Scientiarum*, no. 36 (1989), pp. 23–41, esp. p. 34.

125. Tokyo-Nagoya Working Group of Physics for 1964 Peking Symposium, "Methodology of Elementary Particle Physics in Japan," typewritten draft, Sakata Archival Library, Nagoya University, p. 2.

126. Herbert Passin, "Modernization and the Japanese Intellectual: Some Comparative Observations," in Marius B. Jansen (ed.), *Changing*

Japanese Attitudes Toward Modernization (Princeton, NJ: Princeton University Press, 1965), pp. 447–87, esp. p. 486.

127. Edwin O. Reischauer, *The United States and Japan*, 3rd edn. (Cambridge, MA: Harvard University Press, 1965), p. 315.

128. These ideas are discussed in S.S. Schweber, *In the Shadow of the Bomb: Bethe, Oppenheimer, and the Moral Responsibility of the Scientist* (Princeton: Princeton University Press, 2000).

129. Frank M. Turner, "Public Science in Britain, 1880–1919," *Isis*, vol. 71, no. 259 (December 1980), pp. 589–608, esp. p. 608.

130. For a discussion of the history and significance of the Group, see Tetu Hirosige, "Soryūshiron Guruupu" ("Elementary Particle Theory Group"), *Shizen*, vol. 15, no. 4 (April 1960), pp. 73–81.

131. For the case of pre–World War II China, see James Reardon-Anderson, *The Study of Change: Chemistry in China, 1840–1949* (Cambridge: Cambridge University Press, 1991).

132. Passin, "Modernization and the Japanese Intellectual," p. 486.

133. Beatty, "Scientific Collaboration, Internationalism, and Diplomacy," p. 214.

CHAPTER 5   THE POLITICS OF PURE SCIENCE: YUKAWA AND TOMONAGA

1. US National Archives and Records Administration, "Yukawa Story: Scope and Cotent Note," Item 53506, Record Group 306, Records of the US Information Agency, 1900–1988, Motion Picture, Sound and Video Records LICON, Special Media Archives Services Division (NWCS-M), National Archives at College Park, MD, USA. Online catalog at: www.archives.gov

2. David E. Kaplan, "US Propaganda Efforts in Postwar Japan," *Japan Policy Research Institute Critique*, vol. 4, no. 1 (February 1997).

3. See, e.g., Hideki Yukawa, trans. L.M. Brown and R. Yoshida, *"Tabibito" (The Traveler)* (Singapore: World Scientific, 1982); Hideki Yukawa, trans. John Bester, *Creativity and Intuition: A Physicist Looks at East and West* (Tokyo: Kodansha International, 1973).

4. Takashi Hayakawa, *Nihon no jōryū shakai to keibatsu (Japan's Social Elite and Matrimonial Influence)* (Tokyo: Kadokawa Shoten, 1983), p. 232.

5. Yasutaka Tanikawa, "Introduction and Biographical Sketch," in Hideki Yukawa, *Hideki Yukawa: Scientific Works* (Tokyo: Iwanami Shoten, 1979), pp. vii–viii.

6. Yukawa, *The Traveler*, p. 154.

7. Yukawa, *The Traveler*, p. 170.

8. Translation of Yukawa Family Record Register, ESS files, GHQ/SCAP records, ESS (B) 11648, National Diet Library, Tokyo.

9. Yukawa, *The Traveler*, p. 181.

10. Yukawa, *The Traveler*, p. 188.

11. Takeo Kuwabara, Ken Inoue, and Michiji Konuma (eds), *Yukawa Hideki* (*Hideki Yukawa*) (Tokyo: Nihon Hōsō Shuppan Kyōkai, 1984), p. 325; Yukawa, *The Traveler*, pp. 196–97.

12. Fumitaka Satō, *Yukawa Hideki ga kangaeta koto* (*What Hideki Yukawa Thought*) (Tokyo: Iwanami Shoten, 1985), p. 95.

13. Tanikawa, "Introduction," pp. vii–viii.

14. Yukawa, *The Traveler*, pp. 1–2, 203.

15. Kuwabara et al., *Hideki Yukawa*, p. 326; Yukawa, *Creativity and Intuition*, pp. 38–39.

16. "Curriculum Vitae of Dr. Yukawa," secret, ESS files, GHQ/SCAP records, ESS (A) 09917, National Diet Library, Tokyo.

17. Michiji Konuma, "Social Aspects of Japanese Particle Physics in the 1950s," in Laurie M. Brown, Max Dresden, and Lillian Hoddeson (eds), *Pions to Quarks: Particle Physics in the 1950s* (Cambridge: Cambridge University Press, 1989), pp. 536–48.

18. Hideki Yukawa, manuscript of address given at the inauguration ceremony of the Renewal Committee for Scientific Organization, ESS files, GHQ/SCAP records, ESS (D) 06174, National Diet Library, Tokyo.

19. Kelly to General Marquat, "Utilization of Japanese Scientists by the United States," September 3, 1948, secret, ESS files, GHQ/SCAP records, National Diet Library, Tokyo.

20. For details, see Michiji Konuma, Chieko Masuzawa, and Yoshio Takada, "Resumption of International Relationship of Japanese Particle Physicists after World War II," *Historia Scientiarum*, no. 36 (1989), pp. 23–41.

21. "Tokyo Scientist Her to Study," *The New York Times*, September 4, 1948, p. 4.

22. H.C. Kelly to Colonel L.K. Bunker, "Award of Nobel Peace Prize to Dr. Hideki Yukawa," November 4, 1949, ESS files, GHQ/SCAP records, National Diet Library, Tokyo. Laurence Bunker was a close military aide of General MacArthur's; "Yukawa to Teach at Columbia," *The New York Times*, April 28, 1949, p. 3.

23. "Nobel Prize to U.S. Chemist; Japanese Physicist a Winner," *The New York Times*, November 4, 1949, p. 1.

24. "Nobel Prize Winner Back: Dr. Yukawa to Give Part of His Award to Japanese School," *The New York Times*, December 30, 1949, p. 21.

25. Konuma, "Social Aspects," p. 544; also Kuwabara et al., *Hideki Yukawa*, p. 327.

26. Satio Hayakawa, *Soryushi kara uchū e: Shizen no fukasa o motomete* (*From Elementary Particles to Space: Searching the Depths of Nature*) (Nagoya: Nagoya University Press, 1994), p. 251.

27. Masao Kobayashi, *Saishō Nakasone Yasuhiro: Naikaku sōri daijin e no sokuseki* (*Prime Minister Yasuhiro Nakasone: The Road to the Prime Ministership*) (Tsu: Ise Shinbunsha, 1985), pp. 703–709, esp. p. 703.

28. Yasuhiro Nakasone, "Kagaku Gijutsu Chō setsuritsu made no omoide" ("Memories up until the Establishment of the Science and Technology Agency"), in Science and Technology Agency, *Kagaku Gijutsu Chō: 30 nen no ayumi* (*The Science and Technology Agency: A 30 Year History*) (Tokyo: Sōzō, 1986), pp. 94–95.

29. *Asahi shinbun*, November 15, 1953. Cited in Tetu Hirosige, *Sengo Nihon no kagaku undō* (*Postwar Japan Science Movement*) (Tokyo: Chūō Kōronsha, 1960), p. 90.

30. Hirosige, *Postwar Japan Science Movement*, p. 89.

31. Jon Halliday and Gavan McCormack, *Japanese Imperialism Today: "Co-Prosperitv in Greater East Asia"* (New York: Monthly Review Press, 1973), pp. 107–108.

32. Mitsutomo Yuasa, *Kagakushi* (*The History of Science*) (Tokyo: Tōyō Keizai Shinpōsha, 1961), p. 318; Hirosige, *Postwar Japan Science Movement*, p. 90. The Research Institute was renamed the Defense Agency Technical Research Headquarters in 1959. By 1960, the budget for the institute had become over 2.1 billion yen.

33. John Welfield, *An Empire in Eclipse: Japan in the Postwar American Alliance System: A Study in the Interaction of Domestic Politics and Foreiqn Policy* (London: Athlone Press, 1988), p. 61.

34. Japan Atomic Industrial Forum, *Nihon no genshiryoku: 15 nen no ayumi, jyō* (*Atomic Energy in Japan: A 15 Year History, Part 1*) (Tokyo: Japan Atomic Industrial Forum, 1971), p. 44.

35. Nakasone, "Memories up until the Establishment of the Science and Technology Agency," pp. 94–95.

36. Japan Atomic Industrial Forum, *Nihon no qenshiryoku*, pp. 187–91.

37. Janet E. Hunter (comp.), *Concise Dictionary of Modern Japanese History* (Berkeley: University of California Press, 1984), pp. 113–14.

38. T.J. Pempel, "The Unbundling of 'Japan, Inc.': The Changing Dynamics of Japanese Policy Formation," *Journal of Japanese Studies*, vol. 13, no. 2 (Summer 1987), pp. 271–306.

39. Kobayashi, *Prime Minister Yasuhiro Nakasone*, pp. 703–709, esp. p. 703.

40. Nakasone, "Memories up until the Establishment of the Science and Technology Agency," pp. 94–95.

41. Chitoshi Yanaga, *Biq Business in Japanese Politics* (New Haven: Yale University Press, 1968), pp. 193–94.

42. Japan Atomic Industrial Forum, *Atomic Enerqv*, p. 55.

43. Clark Goodman, "Japan Speeds Organization of Nuclear Program," *Nucleonics* (September 1956), pp. 104–107.

44. See Michael Eckert, "Primacy Doomed to Failure: Heisenberg's Role as Scientific Adviser for Nuclear Policy in the FRG," *Historical Studies in the Physical and Biological Sciences*, vol. 21, part 1 (1990), pp. 29–58.

45. "Yoroku" ("Off the Record"), *Mainichi shinbun*, March 20, 1957, scrapbook, Nobuko Sakata collection.

46. Goodman, "Japan Speeds," pp. 104–107.
47. He was born in 1885. British Government, Intelligence Division of the Far Eastern Bureau, "Who's Who in Japan and Japanese Occupied Territories," confidential report (New Delhi: British Ministry of Information, December 31, 1944), p. 116.
48. Theodore Cohen, ed. by Herbert Passin, *Remaking Japan: The American Occupation as New Deal* (New York: Free Press, 1987), pp. 241–43.
49. Edward Uhlan and Dana L. Thomas, with foreword by Bob Considine, *Shoriki: Miracle Man of Japan, A Biography* (New York: Exposition Press, 1957), pp. 180–82, 196, 202. Also see Ralph Hewins, *The Japanese Miracle Men* (London: Secker and Warburg, 1967), pp. 441–69; Cohen, *Remaking Japan*, pp. 241–43.
50. "Yukawa shi wa naze jihyō dashita?" ("Why did Yukawa hand in his resignation?"), *Asahi shinbun*, March 19, 1957.
51. Japan Atomic Industrial Forum, *Atomic Energy*, pp. 58–59.
52. Japan Atomic Industrial Forum, *Atomic Energy*, pp. 61–62.
53. "Why did Yukawa hand in his resignation?"
54. "Genshiryoku iinkai to Yukawa hakase" ("The Atomic Energy Commission and Dr. Yukawa"), newspaper clipping, late April, 1956, scrapbook, Nobuko Sakata collection.
55. "Why did Yukawa hand in his resignation?"
56. "The Atomic Energy Commission and Dr. Yukawa."
57. Foster Hailey, "Japan Divided on Atomic Plans," *The New York Times*, January 8, 1956, p. 21.
58. "Why did Yukawa hand in his resignation?"
59. Yūichi Yuasa, "Genshiryoku iin jinin shitai: Yukawa hakase no itsuwaranu shinkyō" ("I wish to resign from the Atomic Energy Commission: Dr. Yukawa's True Feelings"), *Mainichi shinbun*, March 21, 1957, scrapbook, Nobuko Sakata collection.
60. "Japanese Scientist Resigns," *The New York Times*, March 19, 1957, p. 39.
61. "Yoroku."
62. For details of Kikuchi's career, see Kikuchi Kinen Jigyōkai Henshū Iinkai (ed.), *Kikuchi Seishi: Gyōseki to tsuisō* (*Seishi Kikuchi: Achievements and Reminiscences*) (Tokyo: Editorial Committee for the Group for the Kikuchi Memorial Project, Institute for Nuclear Study, Tokyo University, 1978).
63. Kuwabara et al., *Hideki Yukawa*, pp. 327–28.
64. His writings have been published as a 15-volume set entitled *Tomonaga Shinichirō: Chosakushū* (*The Collected Works of Sin-itirō Tomonaga*) (Tokyo: Misuzu Shobō, ca. 1984, 1985).
65. Masukata Oheda, "Sin-itiro and I," in Makinosuke Matsui and Hiroshi Ezawa (eds), Cheryl Fujimoto and Takako Sano (trans.), *Sin-itiro Tomonaga: Life of a Japanese Physicist* (Tokyo: MY, 1995), pp. 34–37.

66. Makinosuke Matsui (ed.), *Kaisō no Tomonaga Sin-itirō* (*Reminiscences of Sin-itirō Tomonaga*) (Tokyo: Misuzu Shobō, 1980), pp. 3–6, 8.
67. Matsui, *Reminiscences*, pp. 3–7, 53, 391–92.
68. Matsui, *Reminiscences*, pp. 137–40, 392; Julian S. Schwinger, *Tomonaga Sin-itirō: A Memorial, Two Shakers of Physics* (Tokyo: Nishina Memorial Foundation, 1980), pp. 2–3; Sin-itirō Tomonaga, "Reminiscences," in T. Miyazima (ed.), *Scientific Papers of Tomonaga, Volume 2* (Tokyo: Misuzu Shobō, 1976), pp. 464–67; Minoru Kobayashi, "Riken jidai no Tomonaga san" ("Tomonaga During His Riken Days"), in Daisuke Itō (ed.), *Tsuisō Tomonaga Sin-itirō* (*Reminiscences of Sin-itirō Tomonaga*) (Tokyo: Chūō Kōronsha, 1981), pp. 67–78.
69. Sin-itirō Tomonaga, *Tomonaga Sin-itirō chosaku-shū, bekkan 2: Nikki, shokan* (*Collected works of Sin-itiro Tomonaga, Supplementary Volume 2: Diary and Letters*) (Tokyo: Misuzu Shobō, 1985), p. 130.
70. Takehiko Takabayashi, "Kagakusha toshite, ningen toshite" ("As Scientist, As Human Being"), in Itō, *Reminiscences of Sin-itirō Tomonaga*, pp. 116–34, esp. pp. 121–22; Matsui, *Reminiscences*, pp. 142, 169.
71. "Biography" [Sin-itirō Tomonaga], *Nobel Lectures, Physics: 1963–1970* (Amsterdam: Elsevier, 1972), pp. 137–39.
72. Minoru Oda, "Bannen no Tomonaga Sin-itirō sensei" ("Sin-itirō Tomonaga during His Later Years"), in Itō, *Reminiscences of Sin-itirō Tomonaga*, pp. 99–115, esp. pp. 109–110.
73. Sin-itirō Tomonaga, "Personal History Statement," GHQ/SCAP records, ESS (B) 11647, National Diet Library, Tokyo.
74. Matsui, *Reminiscences*, pp. 217–19.
75. "Japan's Hidden Physicists: Men Who Will Follow in Yukawa's Foosteps," *Nippon Times*, November 22, 1949, GHQ/SCAP records, ESS (E) 06365, National Diet Library, Tokyo.
76. Matsui, *Reminiscences*, pp. 392–94; "Biography," pp. 137–39.
77. Matsui, "From Research Scholar to Scientific Administrator," *Life of a Japanese Physicist*, pp. 203–13, esp. p. 203.
78. Satio Hayakawa, "Kenkyū shinkō ni tsutometa Tomonaga sensei" ("Prof. Tomonaga: A Man Who Strived for the Advancement of Research"), *Kagaku* (*Science*), vol. 49, no. 12 (December 1979), pp. 799–803, esp. p. 799.
79. Konuma, "Social Aspects," pp. 544–45.
80. Matsui, *Reminiscences*, p. 394.
81. Hayakawa, "Prof. Tomonaga," p. 801.
82. For details of its history, see Institute for Nuclear Study, Tokyo University, *Kakken nijū nen shi* (*The Twenty Year History of the Institute for Nuclear Study*) (Tanashi, Tokyo: Institute for Nuclear Study, 1978).
83. Japanese Physics Committee for 1964 Peking Symposium, Working Group of Experimental Nuclear Physics, "The Establishment of the

Institute for Nuclear Study at Tokyo University and Its Results," *Papers Presented at the 1964 Peking Symposium: Natural Science, Vol. 1* (Peking: Scientific and Technical Association of the People's Republic of China, 1965), pp. 311–37, esp. pp. 311–12.

84. Japanese Physics Committee, pp. 312–13.
85. Japanese Physics Committee, pp. 313–14.
86. Goodman, "Japan Speeds," p. 106; information pamphlet, Institute for Nuclear Study, University of Tokyo, 1984; K. Nishimura and F. Sakata (eds), *Introducing INS* (Tokyo: Institute for Nuclear Study, 1985).
87. Japanese Physics Committee, "The Establishment," pp. 314–16.
88. Japanese Physics Committee, pp. 336–37.
89. University-affiliated research institutes of a joint-use nature were established one after the other after INS: Institute for Solid State Physics, Tokyo University (1957); Institute for Protein Research, Osaka University (1958); Institute of Plasma Physics (1961); and the Institute of Space and Aeronautical (later Astronautical) Science, Tokyo University (1964).
90. Nakasone, "Memories up until the Establishment of the Science and Technology Agency," pp. 94–95.
91. Chitoshi Yanaga, *Big Business in Japanese Politics* (New Haven: Yale University Press, 1968), pp. 195–97.
92. Kobayashi, *Prime Minister Yasuhiro Nakasone*, pp. 703–709, esp. p. 704.
93. Kobayashi, *Prime Minister*, p. 704.
94. Kobayashi, *Prime Minister*, p. 704.
95. Nathaniel B. Thayer, *How the Conservatives Rule Japan* (Princeton: Princeton University Press, 1969, 1973), p. 25.
96. Michio Muramatsu and Ellis S. Krauss, "Bureaucrats and Politicians in Policymaking: The Case of Japan," *The American Political Science Review*, vol. 78 (1984), pp. 126–46, esp. p. 143.
97. Matsui, *Reminiscences*, pp. 274–75.
98. Yutaka Osada, "Japanese Policy Attitudes to Antarctica: Mineral Issues," in R.A. Herr and B.W. Davis (eds), *Asia in Antarctica* (Canberra: Centre for Resource and Environmental Studies, Australian National University, 1994), pp. 85–92.
99. Matsui, *Reminiscences*, p. 275; Hayakawa, "Prof. Tomonaga," p. 802.
100. Minutes of 1964 Budget Policy Sub-committee meeting, March 11, 1964, SCNR (February–March 1964) file, Sakata Archives.
101. Reference material attached to minutes of 1964 Budget Policy Sub-committee Meeting, including minutes of ca. March 1964 meeting of Sub-committee for Research Organization, SCNR (February–March 1964) file, Sakata Archives.
102. Report of the Sub-committee for Research Organization, contained in minutes of the March 27, 1964 SCNR meeting, SCNR (February–March 1964) file, Sakata Archives.

103. For a study of his contribution to quantum electrodynamics, see Silvan S. Schweber, *QED and the Men Who Made It: Dyson, Feynman, Schwinger, and Tomonaga* (Princeton: Princeton University Press, 1994), chapter 6.
104. Matsui, *Reminiscences*, p. 331.
105. Matsui, *Reminiscences*, pp. 331, 395–97.
106. Takabayashi, "As Scientist, As Human Being," p. 117.
107. Hideki Yukawa, "The Oriental Approach" (1948), reprinted in Yukawa, *Creativity and Intuition*, pp. 51–60, esp. p. 56.
108. Hideki Yukawa, "Modern Trend of Western Civilization and Cultural Peculiarities in Japan," in C.A. Moore (ed.), *Philosophy and Culture: East and West* (Honolulu: University of Hawaii Press, 1962), pp. 188–98; Yasutaka Tanikawa, "Introduction and Biographical Sketch," in Yukawa, *Scientific Works*, p. ix.
109. Yukawa, *The Traveler*, p. 203.
110. Hideki Yukawa, "A Hundred Years of Science in Japan," in Yukawa, *Scientific Works*, p. 464.
111. Gino K. Piovesana, *Contemporary Japanese Philosophical Thought* (New York: St. John's University Press), pp. 79–109.
112. Mituo Taketani, *Shisō o oru (The Interweaving of Ideas)* (Tokyo: Asahi Shinbunsha, 1985), pp. 154–57.
113. Takabayashi, "As Scientist, As Human Being," pp. 118–19.
114. Takabayashi, "As Scientist, As Human Being," p. 129.
115. Takabayashi, "As Scientist, As Human Being," pp. 129–30.
116. Seitarō Nakamura, *Yukawa Hideki to Tomonaga Shin-ichirō (Hideki Yukawa and Sin-itirō Tomonaga)* (Tokyo: Yomiuri Shinbunsha, 1992).
117. Edwin O. Reischauer, *The United States and Japan* (Cambridge, MA: Harvard University Press, 1965), p. 311.
118. Shigeru Nakayama, "Kagakusha Kyōto Kaigi" ("Kyoto Conference of Scientists"), *Rekishi to shakai (History and Society)*, no. 6 (June 1985), pp. 144–68, esp. pp. 149–50.
119. Nakayama, "Kyoto Conference of Scientists," pp. 150–51.
120. Satō, *What Hideki Yukawa Thought*, p. 162.
121. Toshiyuki Toyoda, "Tsumi o shitta gendai butsurigaku" ("The Modern Physics Which Has Known Sin"), in Itō, *Reminiscences of Sin-itirō Tomonaga*, pp. 135–62, esp. p. 149.
122. Yukawa, *Creativity and Intuition*, p. 191.
123. Yukawa, *The Traveler*, p. 64.
124. Hitoshi Yoshioka, *Kagakusha wa kawaru ka: Kagaku to shakai no shisō shi (Will Scientists Change?: An Intellectual History of Science and Society)* (Tokyo: Shakai Shisōsha, 1984), pp. 169–70.
125. Nakayama, "Kyoto Conference," pp. 151–57, 164.
126. Yukawa, *Creativity and Intuition*, pp. 201–202.
127. Toshiyuki Toyoda, "Scientists Look at Peace and Security," *Bulletin of the Atomic Scientists*, vol. 40, no. 2 (February 1984), pp. 16–19.
128. Kuwabara et al., *Hideki Yukawa*, p. 329.

129. Nakayama, "Kyoto Conference," pp. 156–59.
130. Nakayama, "Kyoto Conference of Scientists," p. 154.
131. See Martin Bulmer, "The Rise of the Academic as Expert," *Minerva*, vol. 25, no. 3 (Autumn 1987), pp. 362–74.

CHAPTER 6   CORPORATE SCIENCE: SAGANE

1. Sagane did not actually find out about this until some two months later, in late September, through one of his students.
2. Ryōkichi Sagane, "Sensō o koete: Genshi bakudan to tomo ni tōka sareta waga tomo no tegami" ("Beyond War: A Letter from Our Friends which was Dropped with the Atomic Bomb"), in Publication Committee (ed.), with a preface by Seiji Kaya, *Sagane Ryōkichi kinen bunshū* (*Collection of Writings to Commemorate Ryōkichi Sagane*) (Tokyo: Sagane Ryōkichi kinen bunshū shuppankai, 1981), pp. 1–11, esp. p. 11.
3. Alwyn McKay, *The Making of the Atomic Age* (Oxford: Oxford University Press, 1984), p. 117; Setsuko Sengoku, "Kaiko dan" ("Recollections") in Publication Committee, pp. 12–16, esp. p. 12.
4. News clipping, December 28, 1949, Terry Hansen, "Japanese Savant Here Tells How Bomb Forced Surrender," *Berkeley Daily Gazette*, Ernest Orlando Lawrence Correspondence and Papers, BANC MSS 72/117C, Carton 9, Folder 39, The Bancroft Library, University of California, Berkeley.
5. General Headquarters, United States Armed Forces, Pacific, Scientific and Technical Advisory Section, "Report on Scientific Intelligence Survey in Japan, September and October 1945, Volume 3" (November 1, 1945), Appendix 5-b-1. [Box 2299, 7/13/44/3, Records of the War Department General and Special Staffs (Record Group 165), Publication File].
6. General Headquarters, United States Armed Forces, Pacific, Scientific and Technical Advisory Section, "Report on Scientific Intelligence Survey in Japan, September and October 1945, Volume 1" (November 1, 1945), p. 16 [Box 2299, 7/13/44/3, Records of the War Department General and Special Staffs (Record Group 165), Publication File].
7. Letter, September 22, 1948, Ryōkichi Sagane to Ernest O. Lawrence, Ernest Orlando Lawrence Correspondence and Papers, BANC MSS 72/117C, Carton 9, Folder 39, The Bancroft Library, University of California, Berkeley.
8. Akira Aizu, "Hito gakusei no omoide" ("The Recollections of One Student"), in Publication Committee, *Ryōkichi Sagane*, pp. 395–98.
9. Translation of extract of Census Register, Ōmura city, Nagasaki prefecture, September 17, 1949; GHQ/SCAP records, ESS (B) 11649, "Travel Abroad of Dr. Sagane" file, p. 9, National Diet Library, Tokyo.

10. Masa Takeuchi, "Nishina kenkyūshitsu shoki no Sagane san" ("Sagane during the Early Days of the Nishina Laboratory"), in Publication Committee, *Ryōkichi Sagane*, pp. 290–94.

11. "Genshikaku kenkyū: Riken jidai o chūshin toshite" ("Nuclear Research: Focusing on the Riken Days"), transcript of discussion, in Publication Committee, pp. 125–56, esp. pp. 127, 132, and p. III.

12. Letter, July 26, 1935, Y. Nishina to E.O. Lawrence, Ernest Orlando Lawrence Correspondence and Papers, BANC MSS 72/117C, Carton 9, Folder 39, The Bancroft Library, University of California, Berkeley; J.L. Heilbron and Robert W. Seidel, *Lawrence and His Laboratory: A History of the Lawrence Berkeley Laboratory, Vol. 1* (Berkeley: University of California Press, 1989), p. 318.

13. Luis W. Alvarez, "Professor Ryokichi Sagane: Personal Recollections," in Publication Committee, *Ryōkichi Sagane*, pp. 415–22.

14. Background resumé of Ryōkichi Sagane, GHQ/SCAP records, ESS (B) 11649, National Diet Library, Tokyo.

15. Kōji Fushimi, "Sagane Ryōkichi sensei no muttsu no katsudō sō" ("Six Aspects of the Activities of Prof. Ryōkichi Sagane"), in Publication Committee, *Ryōkichi Sagane*, pp. 121–24, esp. p. 121, and p. III.

16. Letter, ca. late 1945, Sagane to Lawrence, in Japanese, accompanied by translation and letter, January 4, 1946, Major General Leslie R. Groves to E.O. Lawrence, Ernest Orlando Lawrence Correspondence and Papers, BANC MSS 72/117C, Carton 8, Folder 7, The Bancroft Library, University of California, Berkeley.

17. Seiji Kaya, "S.L. no koto" ("On the Topic of Scientific Liaison") in Publication Committee, *Ryōkichi Sagane*, pp. 275–77, esp. p. 275, and p. III.

18. Masao Kotani, "Sagane kun o shinobu" ("Recollections of Sagane"), in Publication Committee, pp. 277–80, esp. p. 278.

19. General Headquarters, "Volume 3" (November 1, 1945), Appendix 5-b-2.

20. Seiji Kaya, "On the Topic of Scientific Liaison," p. 276.

21. "Summary of Activities of Scientific and Technical Division in Science Reorganization," draft, pp. 3–4, GHQ/SCAP records, ESS (B) 11598, National Diet Library, Tokyo.

22. Physical Society of Japan (ed.), *Nihon no butsurigaku shi: Ge, shiryō hen* (*The History of Physics in Japan: Part B, Source Materials*) (Tokyo: Tokai University Press, 1978), pp. 399–405, esp. p. 405.

23. Japan Atomic Industrial Forum, *Nihon no genshiryoku: 15 nen no ayumi, jyō* (*Atomic Energy in Japan: A 15 Year History, Part 1*) (Tokyo: Japan Atomic Industrial Forum, 1971), pp. 22–24.

24. Letter, March 10, 1947, Sagane to Lawrence, Ernest Orlando Lawrence Correspondence and Papers, BANC MSS 72/117C, Carton 9, Folder 39, The Bancroft Library, University of California, Berkeley.

25. Letter, July 12, 1949, H.C. Kelly to G.W. Fox, GHQ/SCAP records, ESS (B) 11649, "Travel Abroad of Dr. Sagane" file, p. 1, National Diet Library, Tokyo.

26. Letter, October 27, 1949, Fox to Kelly, GHQ/SCAP records, ESS (B) 11649, National Diet Library, Tokyo.
27. Sengoku, "Recollections," p. 15.
28. Alvarez, "Professor Ryokichi Sagane," p. 420.
29. Letter, May 28, 1951, E.O. Lawrence to H. Rowan Gaither, Jr., Associate Director, Ford Foundation, Ernest Orlando Lawrence Correspondence and Papers, BANC MSS 72/117C, Carton 7, Folder 16, The Bancroft Library, University of California, Berkeley.
30. Newspaper clipping, *Asahi shinbun*, February 3, 1948, and English translation, GHQ/SCAP records, ESS (E) 06394, National Diet Library, Tokyo.
31. Chitoshi Yanaga, *Big Business in Japanese Politics* (New Haven: Yale University Press, 1968), p. 178.
32. Yanaga, *Big Business*, pp. 178–79.
33. David E. Lilenthal, *Change, Hope, and the Bomb* (Princeton, NJ: Princeton University Press, 1963), pp. 109–110. Cited in Stephen Hilgartner, Richard C. Bell, and Rory O'Connor, *Nukespeak: Nuclear Language, Visions, and Mindset* (San Francisco: Sierra Club Books, 1982), p. 239.
34. Japan Atomic Industrial Forum, *Atomic Energy*, p. 6.
35. Laura Elizabeth Hein, "Energy and Economic Policy in Postwar Japan," unpublished PhD dissertation, University of Wisconsin, Madison, 1986, pp. 470–71.
36. Japan Atomic Industrial Forum, *Atomic Energy*, pp. 10–11.
37. Japan Atomic Industrial Forum, *Atomic Energy*, p. 11.
38. Peter DeLeon, "Comparative Technology and Public Policy: The Development of the Nuclear Power Reactor in Six Nations," *Policy Sciences*, vol. 11, no. 3 (February 1980), pp. 285–307, esp. p. 297.
39. Japan Atomic Industrial Forum, p. 5; "Dōryoku ro no dōnyū: Genden jidai o chushin to shite" ("The Introduction of a Power Reactor: The Japan Atomic Power Company Days"), roundtable discussion on September 10, 1980, transcript in *Sagane Ryōkichi kinen bunshū* (*Collection of Writings to Commemorate Ryōkichi Sagane*) (Tokyo: Sagane Ryōkichi Kinen Bunshū Shuppankai, 1981), pp. 253–73, esp. p. 256.
40. Masanori Ōnuma, Yōichirō Fujii, and Kunioki Katō, *Sengo Nihon kagakusha undō shi: Jyō* (*A History of the Postwar Japanese Scientists' Movement: Part One*) (Tokyo: Aoki Shoten, 1975), p. 120.
41. Seiji Kaya, "The Activities Shown by Dr. Fujioka at the Initial Stage of the Peaceful Use of Atomic Energy in Japan," *Yoshio Fujioka Commemorative Issue* (Tokyo: The Institute for optical Research, Tokyo University of Education, 1967), pp. 370–72, esp. p. 371.
42. Yanaga, *Big Business*, pp. 179–80.
43. Kaya, "The Activities Shown by Dr. Fujioka," pp. 370–72, esp. p. 371.
44. Kaya, "The Activities Shown by Dr. Fujioka," p. 371.

45. Masao Kobayashi, *Saishō Nakasone Yasuhiro: Naikaku sōri daijin e no sokuseki* (*Prime Minister Yasuhiro Nakasone: The Road to the Prime Ministership*) (Tsu: Ise Shinbunsha, 1985), pp. 703–709, esp. p. 703.

46. Janet E. Hunter (comp.), *Concise Dictionary of Modern Japanese History* (Berkeley: University of California Press, 1984), p. 56.

47. See Aurelia George, "The Nakasone Challenge: Historical Constraints and New Initiatives in Japan's Defence Policy," *Legislative Research Service Discussion Paper*, no. 2 (1986–1987), pp. 24–25.

48. Science Council of Japan, *Annual Report, 1951–1958* (Tokyo, 1959), p. 110.

49. Japan Atomic Industrial Forum, *Atomic Energy*, p. 44.

50. Japan Atomic Industrial Forum, *Atomic Energy*, p. 16. For an account of the incident, see Ralph Lapp, *The Voyage of the Lucky Dragon* (New York: Harper and Row, 1958).

51. Yanaga, *Big Business*, p. 181.

52. Hein, "Energy and Economic Policy," pp. 471, 473–74.

53. Stanley Levey, "Nuclear Reactor Urged for Japan," *New York Times*, September 22, 1954, p. 14, cited in Hein, pp. 470–71, 490.

54. Hein, "Energy and Economic Policy," pp. 470–71.

55. See Kōji Fushimi, "Genshiryoku heiwa kōsei ni dō taishō suru ka" ("How to Deal with the Atomic Power Peace Offensive?"), *Chūō kōron*, vol. 70, no. 6 (June 1955), pp. 34–42.

56. "Genshiryoku kenkyū kaihatsu: Genken jidai o chūshin toshite" ("The Development of Nuclear Power: Focusing on the JAERI Days"), transcript of discussion, in Publication Committee, *Ryōkichi Sagane*, pp. 219–34, esp. p. 220.

57. Letter, August 3, 1955, A.W. Hand, Professional Personnel, to A. Kuckern, Chief, Entry and Departure Section, US Immigration and Naturalization Service, San Francisco, Ernest Orlando Lawrence Correspondence and Papers, BANC MSS 72/117C, Carton 9, Folder 39, The Bancroft Library, University of California, Berkeley.

58. Letter, ca. late 1956, Sagane to Lawrence, Ernest Orlando Lawrence Correspondence and Papers, BANC MSS 72/117C, Carton 9, Folder 39, The Bancroft Library, University of California, Berkeley. On Randolph Hotel, Oxford letterhead, sent from England while visiting as a member of the Japanese government atomic energy delegation.

59. Hafstad's name is misspelt in Yanaga, *Big Business*, p. 185.

60. "Lawrence R. Hafstad," *Memorial Tributes: National Academy of Engineering*, vol. 7 (1994), pp. 104–107.

61. "Use Atom Power, Industry is Urged: Hafstad of AEC Charges Timidity and Warns Against Defaulting to Russia," *The New York Times*, September 29, 1954, p. 16.

62. Letter, January 10, 1956, Sagane to Lawrence, Ernest Orlando Lawrence Correspondence and Papers, BANC MSS 72/117C, Carton 9, Folder 39, The Bancroft Library, University of California, Berkeley.

63. "Notes on Dr. Ryokichi Sagane's visit with D. Cooksey," December 17, 1956, Ernest Orlando Lawrence Correspondence and Papers, BANC MSS 72/117C, Carton 9, Folder 39, The Bancroft Library, University of California, Berkeley.

64. Tamaki Ipponmatsu, "Sagane hakase no omoide" ("Recollections of Dr. Sagane") and Hiroshi Murata, "Sagane sensei o shinobu" ("Recollections of Prof. Sagane") in Publication Committee, *Ryōkichi Sagane*, pp. 330–33, 333–35.

65. "The Development of Nuclear Power," p. 225.

66. "The Development of Nuclear Power," p. 221.

67. R. Imai, "Japan and the Nuclear Age," *Bulletin of the Atomic Scientists*, vol. 26, no. 6 (June 1970), p. 37.

68. Lawrence Olson, "Atomic Cross-Currents in Japan," *American Universities Field Staff Reports: East Asia Series*, vol. 7, no. 5 (April 1959), p. 4.

69. National Academy of Sciences, *Federal Support of Basic Research in Institutions of Higher Learning* (Washington, DC: National Research Council, 1964), pp. 16–56, reprinted in James L. Penick, Jr., Carroll W. Pursell, Jr., Morgan B. Sherwood, and Donald C. Swain (eds), *The Politics of American Science, 1939 to the Present* (Cambridge, MA: MIT Press, 1972 ed.), pp. 2–41, esp. p. 27.

70. Charles F. Bingman, *Japanese Government Leadership and Management* (Houndmills, Basingstoke, Hampshire: Macmillan Press, 1989), p. 122.

71. Prime Minister of Japan and His Cabinet, "The National Administrative Organization in Japan," online at *http://www.kantei.go.jp/*. Viewed October 25, 2004.

72. Japan Atomic Industrial Forum, *Atomic Energy*, p. 222.

73. Japan Atomic Energy Research Institute, "Description of the Japan Atomic Energy Research Institute" (August 1, 1956), p. 3. Details of the history of the establishment of the Institute can be found in Ichirō Nakajima (ed.), *Zaidan hōjin Genshiryoku Kenkyūjoshi* (*A History of the Japan Atomic Energy Research Institute*) (Tokyo: Japan Atomic Energy Research Institute, March 1957).

74. "The Development of Nuclear Power," p. 220.

75. J.E. Hodgetts, *Administering the Atom for Peace* (New York: Atherton Press, 1964), pp. 110–12; Hiroshi Murata, "Nuclear Energy: The Next 10 Years," *New Scientist*, vol. 36, no. 571 (November 16, 1967), Japanese Supplement, pp. 3–4.

76. "US Concern to Build Nuclear Reactor for Japan," *The New York Times*, December 5, 1956, p. 59.

77. Publication Committee, *Ryōkichi Sagane*, p. IV.

78. JRR-2 (a heavy-water reactor) used American technology as well, but was mainly manufactured by the Japanese. It went critical in 1962, followed by the Japan-made JRR-3 (a 10-MWe natural uranium, heavy-water reactor), which went critical two years later. See *Japanese*

Industry (Tokyo: Foreign Capital Research Society, Bank of Japan, ca. 1960); Olson, "Atomic Cross-Currents in Japan," Motohisa Mori, *Genshiryoku nenpyō (1934–1985) (Chronological History of Atomic Energy)* (Tokyo: Marunouchi Shuppan, 1986).

79. Publication Committee, *Ryōkichi Sagane*, p. IV.
80. Mitutomo Yuasa, *Kagakushi (The History of Science)* (Tokyo: Tōyō Keizai Shinpōsha, 1961), pp. 334–35.
81. Yuasa, *History of Science*, p. 334–36.
82. Tetu Hirosige, *Sengo Nihon no kagaku undō (Japan's Postwar Science Movement)* (Tokyo: Chūō Kōronsha, 1960), p. 121.
83. Publication Committee, *Ryōkichi Sagane*, p. IV.
84. Publication Committee, *Ryōkichi Sagane*, p. IV.
85. Actual construction commenced in 1967 at the dockyard of Ishikawajima-Harima Heavy Industry Company. While it was hoped that the vessel of 8,300 gross tons would be used partly as a cargo ship, rising costs and technical problems have continued to plague its development and have effectively ruled out its use for commercial purposes. From the time of its first voyage, the ship began to leak radioactive materials. Murata, "Nuclear Energy: The Next 10 Years." By 1992, JAERI's "Mutsu" had completed its mission as a nuclear powered experimental ship and plans were underway for its decommission. *STA Today*, vol. 4, no. 2 (February 1992), p. 6.
86. Japan Atomic Industrial Forum, *Atomic Energy*, p. 131.
87. Japan Atomic Industrial Forum, *Atomic Energy*, p. 135.
88. Hiroshi Murata, "Sagane sensei o shinobu" ("Recollections of Prof. Sagane"), in Publication Committee, *Ryōkichi Sagane*, pp. 333–35.
89. Masahiro Yoneyama, "Nuclear Reactor Still Hot Potato," *Asahi Evening News*, November 9, 1988; *Japanese Industry* (Tokyo: Foreign Capital Research Society, Bank of Japan, ca. 1962), pp. 39–40.
90. Tatsujirō Suzuki and Atsuyuki Suzuki, "Japan's Nuclear Energy Policy," *Science and Public Policy*, vol. 13, no. 1 (February 1986), pp. 25–32.
91. Seiichi Takahata, *Industrial Japan and Industrious Japanese* (Osaka: Nissho Co., 1968), pp. 144–74; Victor Gilinsky and Paul Langer, *The Japanese Civilian Nuclear Program*, RAND Memorandum RM-5366-PR (Santa Monica: RAND, August 1967), p. 15.
92. Publication Committee, *Ryōkichi Sagane*, p. IV.
93. "Japan Atomic Power Co. Celebrates 15th Anniversary," *Atoms in Japan*, no. 12 (December 1972), pp. 24–29, courtesy of Takamichi Mito.
94. Rodney L. Huff, "Political Decisionmaking in the Japanese Civilian Atomic Energy Program," unpublished PhD dissertation, George Washington University, 1973, pp. 45–46.
95. R.P. Suttmeier, "The Japanese Nuclear Power Option: Technological Promise and Social Limitations," in R.A. Morse (ed.), *The Politics of Japan's Energy Strategy* (Berkeley: Institute of East Asian Studies, University of California, Berkeley, 1981), pp. 106–166, esp. p. 110.

96. J.E. Endicott, *Japan's Nuclear Option* (New York: Praeger, 1975), p. 87.
97. "Kaku yūgō kenkyū no suishin" ("The Promotion of Nuclear Fusion"), in Publication Committee, *Ryōkichi Sagane*, pp. 235–52, esp. p. 236.
98. Fushimi, "Six Aspects of the Activities of Prof. Ryōkichi Sagane," p. 124; and "The Promotion of Nuclear Fusion," p. 237.
99. "The Promotion of Nuclear Fusion," p. 240; and Ryōji Ueda, "Sagane sensei no omoide" ("Recollections of Prof. Sagane"), in Publication Committee, *Ryōkichi Sagane*, pp. 280–83.
100. Fushimi, "Six Aspects," p. 124; and "The Promotion of Nuclear Fusion," p. 237.
101. The idea of "policy cultures" with different political and social interests, different institutional bases, and traditions is developed in Sheila Jasanoff, Gerald E. Markle, James C. Petersen, and Trevor Pinch (eds), *Handbook of Science and Technology Studies* (Thousand Oaks, CA: Sage Publications, 1995), esp. pp. 575–76.
102. Herbert Passin, "Intellectuals in the Decision Making Process," in Ezra F. Vogel (ed.), *Modern Japanese Organization and Decision-making* (Berkeley: University of California Press, 1975), pp. 251–83.
103. Spoken at meeting of Japan Committee for Economic Development and reported in *Nihon keizai shinbun*, December 25, 1961. Cited in Yanaga, *Big Business*, p. 34.
104. For a discussion of the transfer of nuclear technology, see International Atomic Energy Agency (IAEA), *Developing Industrial Infrastructures to Support a Programme of Nuclear Power: A Guidebook* (Vienna: IAEA, 1988).

CHAPTER 7 SCIENCE ON THE INTERNATIONAL STAGE: HAYAKAWA

1. Satio Hayakawa, "Genbaku chōsa ni o tomo shite" ("Accompanying the Atomic Bomb Survey"), in Publication Committee (ed.), with a preface by Seiji Kaya, *Sagane Ryōkichi kinen bunshū* (*Collection of Writings to Commemorate Ryōkichi Sagane*) (Tokyo: Sagane Ryōkichi Kinen Bunshū Shuppankai, 1981), pp. 284–86, esp. p. 284.
2. Nobuyuki Fukuda, "Kuchi, te, ashi, mina hatchō" ("Mouth, hands, legs, all skillful"), *Asahi jaanaru*, March 26, 1961, p. 31.
3. Nagoya University, *Hayakawa Sachio sensei tsuisōroku* (*Collection of Reminiscences of Professor Satio Hayakawa*) (Department of Astrophysics, Nagoya University, September 1994), p. 25.
4. Kiyonobu Itakura, "Kagakusha no jishuteki na soshiki" ("An Independent Organization of Scientists"), in Mituo Taketani (ed.), *Shizen kagaku gairon, daiikkan: Kagaku gijutsu to Nihon shakai*

(*An Introduction to the Natural Sciences, Vol. 1: Science, Technology and Japanese Society*) (Tokyo: Keisō Shobō, 1962), pp. 155–73.

5. Fukuda, "Mouth, hands, legs, all skillful," p. 31.

6. Satio Hayakawa and Mamoru Saitō, "Astronomy in Japan," *Astrophysics and Space Science*, vol. 99 (1984), pp. 393–402, esp. p. 401.

7. John W. Dower, "The Useful War," *Daedalus*, vol. 119, no. 3 (Summer 1990), pp. 49–70, esp. p. 56; Ben-Ami Shillony, *Politics and Culture in Wartime Japan* (Oxford: Clarendon Press, 1991), p. 139; Laurie M. Brown and Yoichiro Nambu, "Physicists in Wartime Japan," *Scientific American*, vol. 279, no. 6 (December 1998), pp. 96–103.

8. "Hayakawa Sachio sensei goryakureki" ("Brief Personal History of Prof. Satio Hayakawa"), Hayakawa collection. Hayakawa was born on October 16, 1923 in Niihama city in Ehime prefecture. In 1930, he entered the Jiyū Gakuen school in Tanashi, Tokyo. The following year, he transferred to the Sumitomo Sōkai Elementary School in Niihama. In 1936, Hayakawa advanced to the Musashi High School.

9. Satio Hayakawa, "List of Academic Publications," Hayakawa collection.

10. Hayakawa, "Accompanying the Atomic Bomb Survey," pp. 284–86, esp. p. 284.

11. Nagoya University, pp. 3–4.

12. Nagoya University, *Collection of Reminiscences*, pp. 3, 88–90.

13. "Hayakawa Sachio: Kinmu kiroku kādo" ("Satio Hayakawa: Work Record"), Hayakawa collection; Satio Hayakawa, "Kenkyū shinkō ni tsutometa Tomonaga sensei" ("Prof. Tomonaga: A Man Who Strived for the Advancement of Research"), *Kagaku* (*Science*), vol. 49, no. 12 (December 1979), pp. 799–803, esp. p. 799.

14. Hayakawa, Prof. Tomonaga, pp. 799–803, esp. p. 799.

15. Personal history of Satio Hayakawa, GHQ/SCAP records, ESS (E) 06383, National Diet Library, Tokyo.

16. "Brief Personal History"; Laurie M. Brown, "Yoichiro Nambu: The First Forty Years," *Progress of Theoretical Physics Supplement*, no. 86 (1986), pp. 1–11, draft of manuscript, p. 5, courtesy of Prof. Laurie Brown, Northwestern University.

17. Bruno Rossi to Hayakawa, April 3, 1950, GHQ/SCAP records, ESS (E) 06384, National Diet Library, Tokyo.

18. "Science and Foreign Relations," US State Department Publication 3860 (May 1950), GHQ/SCAP records, ESS (B) 11758, National Diet Library, Tokyo.

19. Satio Hayakawa, "Galactic Origin of Cosmic Rays from Fermi's Theory to Gamma-Ray Astronomy," in Giacomo Cavallo, Shuji Fukui, Hideyuki Matsumura, and Toshiyuki Toyoda (eds), *Creativity and Inspiration: Perspectives of Collaboration in Mathematics and Physics between Italy and Japan* (December 1987, publisher unknown), pp. 65–77, esp. p. 65, Hayakawa collection.

20. "Brief Personal History."
21. Michiji Konuma, "Social Aspects of Japanese Particle Physics in the 1950s," in Laurie M. Brown, Max Dresden, and Lillian Hoddeson (eds), *Pions to Quarks: Particle Physics in the 1950s* (Cambridge: Cambridge University Press, 1989), pp. 536–48, esp. pp. 540–41; Satio Hayakawa, "Kyōdō riyō kenkyūjo: Kako, genzai, mirai" ("Joint-use Research Institutes: Past, Present and Future"), *Pariti (Parity)*, vol. 3, no. 1 (1988), pp. 80–83, esp. p. 80.
22. Minoru Oda, "Italian Influence on Progress of Space Science in Japan: Personal Memoir," in Cavallo et al., *Creativity and Inspiration*, pp. 71–77.
23. "Satio Hayakawa: Work Record."
24. Masao Kotani, "Dr. Y. Fujioka and the Science Council of Japan," in *The Yoshio Fujioka Commemorative Issue* (Tokyo: Institute for Optical Research, Tokyo University of Education, 1967), pp. 353–56, esp. p. 354.
25. Richard P. Feynman, as told to Ralph Leighton, edited by Edward Hutchings, *Surely You're Joking, Mr. Feynman!: Adventures of a Curious Character* (New York: Norton and Co., 1985), p. 237.
26. Feynman, *Surely You're Joking*, p. 238.
27. Feynman, *Surely You're Joking*, p. 240.
28. Feynman, *Surely You're Joking*, p. 238.
29. Feynman, *Surely You're Joking*, p. 241.
30. Feynman, *Surely You're Joking*, p. 240.
31. Seitarō Nakamura, *Watakushi no ayunda michi: Yukawa chūkanshi to tomo ni (The Path Which I Have Followed: With Yukawa's Meson* (Tokyo: Tōkai Daigaku Shuppankai, 1991), pp. 18–19.
32. "Hayakawa Sachio kyōju no gyōseki ni tsuite" ("On the Achievements of Prof. Satio Hayakawa"), Hayakawa collection.
33. "On the Achievements of Prof. Satio Hayakawa."
34. "Satio Hayakawa: Work Record."
35. Fukuda, "Mouth, hands, legs, all skillful," p. 31.
36. Fumitaka Satō, "Hayakawa sensei no omoide" ("Memories of Professor Hayakawa"), in Satio Hayakawa, *Soryushi kara uchū e: Shizen no fukasa o motomete (From Elementary Particles to Space: Searching the Depths of Nature)* (Nagoya: Nagoya University Press, 1994), insert, pp. 5–7, esp. p. 5.
37. During Feynman's visit to Japan in 1955, his tendency to ask difficult questions was referred by the Japanese as "Feynman's Bombardments." Feynman, *Surely You're Joking, Mr. Feynman!*, p. 245.
38. S. Hayakawa, "The Origin of Cosmic Rays," in S. Hayakawa, H.Y. Chiu, G. Feinberg, and M. Dresden, *Lectures on Astrophysics and Weak Interactions* (New York: Gordon and Breach, 1964), pp. 1–164.
39. Satio Hayakawa, *Cosmic Ray Physics: Nuclear and Astrophysical Aspects* (New York: Wiley-Interscience, 1969).

40. "Kōseki chōsho: Hayakawa Sachio" ("Record of Meritorious Service: Satio Hayakawa"); "Brief Personal History."

41. These publications include: Laurie M. Brown, Michiji Konuma, and Zirō Maki (eds), *Particle Physics in Japan, 1930–1950*, 2 vols (Kyoto: Research Institute for Fundamental Physics, 1980); Laurie M. Brown, Rokuo Kawabe, Michiji Konuma, and Zirō Maki (eds), *Elementary Particle Theory in Japan, 1935–1960: Japan–USA Collaboration, Second Phase* (Kyoto: Yukawa Hall Archival Library, Research Institute for Fundamental Physics, 1988); Laurie M. Brown, Rokuo Kawabe, Michiji Konuma, and Zirō Maki (eds), "Elementary Particle Theory in Japan, 1930–1960: Proceedings of the Japan–USA Collaborative Workshops," *Progress of Theoretical Physics Supplement*, no. 105 (1991), special issue.

42. "Tsune ni shinkenkyū no sentō" ("Always at the Forefront of New Research"), *Asahi shinbun*, February 6, 1992; Satio Hayakawa, "Supernova Origin of Cosmic Rays," *Progress of Theoretical Physics*, vol. 15 (1956), pp. 111–21.

43. "Record of Meritorious Service"; "Brief Personal History."

44. Peter J. Westwick, *The National Labs: Science in an American System, 1947–1974* (Cambridge, MA: Harvard University Press, 2003).

45. Joan Lisa Bromberg, *Fusion: Science, Politics, and the Invention of a New Energy Source* (Cambridge, MA: MIT Press, 1982), pp. 32–34.

46. Bromberg, *Fusion*, pp. 6–7.

47. *Proceedings of the International Conference on the Peaceful Uses of Atomic Energy, held in Geneva, 8 August–20 August 1955, Vol. 16: Record of the Conference* (New York: United Nations, 1956), pp. 141–42.

48. Reported in *The New York Times*, August 9, 1955, p. 1. Cited by Bromberg, *Fusion*, p. 67. See also H. Bhabha, "Presidential Address," *Proceedings*, p. 35.

49. Satio Hayakawa and Kazue Kimura, "Kakuyūgō kenkyū kotohajime (1)" ("The Beginnings of Nuclear Fusion Research: Part One"), *Kakuyūgō kenkyū*, vol. 57, no. 4 (April 1987), pp. 201–214, esp. p. 201.

50. Nagoya University, Institute of Plasma Physics, *Purazuma Kenkyūjo 10-nen no ayumi* (*The Ten Year History of the Institute of Plasma Physics*) (Nayoya, 1972), pp. 1–2.

51. Tetu Hirosige, "Kakuyūgō hannō to wa: Rekishi to tenbō" ("What is Nuclear Fusion? Its History and Development"), *Shizen*, vol. 13, no. 5 (May 1958), pp. 3–12, esp. p. 12.

52. Richard G. Hewlett and Jack M. Holl, *Atoms for Peace and War, 1953–1961: Eisenhower and the Atomic Energy Commission* (Berkeley: University of California Press, 1989), pp. 525–27.

53. Bromberg, *Fusion*, p. 93.

54. Hewlett and Holl, *Atoms for Peace*, pp. 525–27.

55. Hayakawa to Yukawa, September 19, 1955, *Soryūshiron kenkyū* (*Studies in Elementary Particle Theory*), vol. 9, no. 5 (1955), pp. 531–32.

Cited in Sigeko Nisio, "Satio Hayakawa's Letter to Hideki Yukawa (September 19, 1955): A Trigger for Controlled Nuclear Fusion Research in Japan," *Historia Scientiarum*, vol. 1, no. 3 (1992), pp. 221–22.
56. Hirosige, "What is Nuclear Fusion?," p. 12.
57. Bromberg, *Fusion*, p. 70.
58. Hirosige, "What is Nuclear Fusion?," p. 12.
59. Hayakawa and Kimura, "Fusion Research: Part One," pp. 202–203.
60. The 1949–1971 rate of exchange was set at 360 yen to the US dollar. Al Alletzhauser, *The House of Nomura* (London: Bloomsbury, 1990), Appendix 6.
61. Hayakawa and Kimura, "Fusion Research: Part One," p. 203.
62. Hayakawa and Kimura, "Fusion Research: Part One," p. 203.
63. Nagoya University, *Ten Year History*, pp. 1–2.
64. Hayakawa and Kimura, "Fusion Research: Part One," p. 203.
65. Nagoya University, *Ten Year History*, pp. 1–2.
66. Hayakawa and Kimura, "Fusion Research: Part One," pp. 204–205.
67. Science Council of Japan, *Annual Report 1959* (Tokyo, 1960), pp. 96–97.
68. Satio Hayakawa and Kazue Kimura, "Kakuyūgō kenkyū kotohajime (2)" ("The Beginnings of Nuclear Fusion Research: Part Two"), *Kakuyūgō kenkyū*, vol. 57, no. 5 (May 1987), pp. 271–79, esp. pp. 271–72.
69. Hayakawa and Kimura, "Fusion Research: Part One," p. 205.
70. Hayakawa and Kimura, "Fusion Research: Part Two," p. 275.
71. Science Council, *Annual Report 1959*, p. 97.
72. Nagoya University, *Ten Year History*, p. 3.
73. Hayakawa and Kimura, "Fusion Research: Part Two," pp. 275–76.
74. Satio Hayakawa and Kazue Kimura, "Kakuyūgō kenkyū koto-hajime (3)" ("The Beginnings of Nuclear Fusion Research: Part Three"), *Kakuyūgō kenkyū*, vol. 57, no. 6 (June 1987), pp. 364–78, esp. p. 364.
75. Bromberg, *Fusion*, pp. 22, 88, 98.
76. Nagoya University, *Ten Year History*, p. 3.
77. Bromberg, *Fusion*, pp. 5, 95–99, 110.
78. Hayakawa and Kimura, "Fusion Research: Part Two," p. 271.
79. Bromberg, *Fusion*, pp. 9, 135–36, 172, 198.
80. "On the Achievements of Prof. Satio Hayakawa."
81. Hayakawa and Saitō, "Astronomy in Japan," pp. 395–97.
82. Takeshi Inagaki, "Rocket Readiness," *Japan Quarterly*, vol. 35, no. 2 (April–June 1988), pp. 146–51, esp. p. 146.
83. Mitutomo Yuasa, *Kagakushi* (*The History of Science*) (Tokyo: Toyō Keizai Shinpōsha, 1961), pp. 315–16.
84. James L. Penick, Jr., Carroll W. Pursell, Jr., Morgan B. Sherwood, and Donald C. Swain (eds), *The Politics of American Science, 1939 to the Present* (Cambridge, MA: MIT Press, 1972 edition), pp. 214–17.

85. Science Council of Japan, *Annual Report, 1951–1958* (Tokyo, 1959), p. 210.
86. See Bruce Hevly, "A Scientific Reconnaissance: The IGY Research Program of the US Naval Research Laboratory," paper presented at *History of Laboratories and Laboratory Science*, British–North American Joint Meeting of the Canadian Society for the History and Philosophy of Science, the British Society for the History of Science, and the History of Science Society, Victoria University at the University of Toronto, Canada, July 26–28, 1992, book of abstracts, pp. 33–34.
87. F. Roy Lockheimer, "The Rising Sun in Space: Aspects of the Japanese Approach to Space Science, Part 1: The University of Tokyo," *American Universities Field Staff Reports*, East Asia Series, vol. 14, no. 1 (Japan), [FRL-1-'67] (January 1967), pp. 1–14, esp. p. 3.
88. Institute of Space and Astronautical Science (ISAS), Editorial Committee for the Thirty Year History of Space Observations, *Uchū kūkan kansoku 30-nenshi nenpyō* (*Thirty Year History and Chronological Table of Space Observation*) (Tokyo, 1987), pp. 3–5.
89. Science Council, *Annual Report, 1951–1958*.
90. Science Council, *Annual Report, 1951–1958*, pp. 210–11.
91. ISAS, *Thirty Year History*, pp. 8, 11.
92. ISAS, *Thirty Year History*, pp. 11, 13.
93. ISAS, *Thirty Year History*, p. 14.
94. Masao Kobayashi, *Saishō Nakasone Yasuhiro: Naikaku sōri daijin e no sokuseki* (*Prime Minister Yasuhiro Nakasone: The Road to the Prime Ministership*) (Tsu: Ise Shinbunsha, 1985), pp. 703–709, esp. p. 704.
95. Hayakawa and Saitō, "Astronomy in Japan," p. 396.
96. ISAS, *Thirty Year History*, pp. 13–14.
97. "Satio Hayakawa: Work Record," p. 7.
98. F. Roy Lockheimer, "The Rising Sun in Space: Aspects of the Japanese Approach to Space Science, Part 2: International Co-operation, Organization, and the Future," *American Universities Field Staff Reports*, East Asia Series, vol. 14, no. 2 (Japan) [FRL-2-'67] (February 1967), pp. 1–26, esp. pp. 9–10.
99. Kobayashi, pp. 703–709, esp. p. 704.
100. Inagaki, "Rocket Readiness," p. 148.
101. Minoru Oda, "Italian Influence on Progress of Space Science in Japan: Personal Memoir," in Cavallo et al., *Creativity and Inspiration*, pp. 71–77.
102. Lockheimer, "Rising Sun, Part 1," p. 2.
103. Lockheimer, "Rising Sun, Part 2," p. 13.
104. Lockheimer, "Rising Sun, Part 2," p. 16.
105. For one account, see Lillian Hoddeson, "Establishing KEK in Japan and Fermilab in the US: Internationalism, Nationalism and High Energy Accelerators," *Social Studies of Science*, vol. 13 (1983), pp. 1–48. Also see Sharon J. Traweek, *Beamtimes and Lifetimes: The World of*

High Energy Physicists (Cambridge, MA: Harvard University Press, 1988); Satio Hayakawa and Morris F. Low, "Science Policy and Politics in Post-war Japan: The Establishment of the KEK High Energy Physics Laboratory," *Annals of Science*, vol. 48 (1991), pp. 207–29.

106. Alun Anderson and John Maddox, "High-energy Physics: Joining the Big League," *Nature*, vol. 305 (September 29, 1983), pp. 379–80.

107. Catherine Lee Westfall, "The First 'Truly National Laboratory': The Birth of Fermilab," unpublished PhD dissertation, 2 vols., Michigan State University, 1988, p. 316.

108. Westfall, "The First 'Truly National Laboratory,' " p. 25.

109. Robert W. Seidel, "Accelerators and National Security: The Evolution of Science Policy for High-Energy Physics, 1947–1967," *History and Technology*, vol. 11, no. 4 (1994), pp. 361–91, esp. pp. 361–62.

110. Alun M. Anderson, *Science and Technology in Japan* (Harlow, Essex: Longman, 1984), p. 168.

CONCLUSION

1. Bob Tadashi Wakabayashi, "Introduction," in Bob Tadashi Wakabayashi (ed.), *Modern Japanese Thought* (Cambridge, UK: Cambridge University Press, 1998), pp. 1–29, esp. p. 22.

2. Andrew E. Barshay, "Postwar Social and Political Thought, 1945–90," in Bob Tadashi Wakabayashi (ed.), *Modern Japanese Thought* (Cambridge, UK: Cambridge University Press, 1998), pp. 273–355, esp. p. 304.

3. Minutes of the Special Delibrative Committee (Kōsa Tokubetsu Iinkai), Meeting No. 10, November 30, 1949, House of Representatives, Japan. Online at *http://kokkai.ndl.go.jp*.

4. Barshay, "Postwar Social and Political Thought," p. 317.

5. See the essays contained in Hideki Yukawa, trans. John Bester, *Creativity and Intuition: A Physicist Looks at East and West* (Tokyo: Kodansha International, 1973). Many of the essays were originally written in the 1960s.

6. S.S. Schweber, *In the Shadow of the Bomb: Oppenheimer, Bethe, and the Moral Responsibility of the Scientist* (Princeton: Princeton University Press, 2000).

7. Sin-itiro Tomonaga and Jiro Osaragi, "The Atomic Age: The World and Japan," first published in Japanese in *Mainichi shinbun*, January 4, 1951. Reprinted in Makinosuke Matsui and Hiroshi Ezawa (eds), Cheryl Fujimoto and Takako Sano (trans.), *Sin-itiro Tomonaga: Life of a Japanese Physicist* (Tokyo: MY, 1995), pp. 214–18, esp. p. 216.

8. Schweber, *In the Shadow of the Bomb*, p. 13.

9. Reiji Yoshida, "Japan Considered Developing Nukes: Nakasone," *The Japan Times*, June 19, 2004, online version.

10. Yoshida; The Associated Press, "Japan Considered Nuke Arsenal," *CBSNews.com*, June 18, 2004; Yasuhiro Nakasone, *Jiseiroku: Rekishi hōtei no hikoku toshite* (*The Meditations: A Defendant in a Court of History*) (Tokyo: Shinchōsha, 2004).

11. Yoshida, "Japan Considered Developing Nukes: Nakasone."

GLOSSARY

bushidō	Way of the Warrior (ideal samurai values)
butsurigaku	physics
chūbosu	middle bosses
daibosu	big bosses
daikan	local official
goyōgakusha	government-patronized scholar
Keidanren	Federation of Economic Organizations
kobosu	little bosses
kōgi	public good
kokka	the state
kokugaku	study of ancient Japanese thought and culture
kyūrigaku	physics
Meiji period	1868–1912
Monbushō	Ministry of Education, Science and Culture
Mushashugyō	warrior training
rakugo	comic stories
Shōwa period	1926–1989
sōgō	general, comprehensive
Taishō period	1912–1926
Tokugawa period	ca. 1603–1868
zaibatsu	family financial/industrial combines
zaikai	the business world

INDEX

accelerator physics, 194–5
see also KEK High Energy Physics
    Laboratory
adoption, 2
Agency of Science and Technology
    (Gijutsuin), 33, 35, 47
Allied Occupation of Japan
    (1945–52), 45–71, 80, 86, 92
Alvarez, Luis W., 143, 146
Arakatsu, Bunsaku, 37, 46
Arisawa, Hiromi, 116–17
arms production, 113
astrophysics, 178
Atomic Bomb Casualty Commission
    (ABCC), 92
Atomic Bomb Project, 36–9
atomic energy, 92
Atoms for Peace, 92, 150–1

Baelz, Erwin, 8
Basic Atomic Energy Law, 94
Berkeley, University of California at,
    27, 143, 149
Bernal, J.D., 76–7, 80, 89, 102
Bohr, Niels, 20–1, 26
boiling water reactors (BWRs),
    163–4
bushidō, 1, 253

Cabinet Planning Board, 32
Calder Hall reactor, 152
China
    see People's Republic of China

Civil Censorship Detachment
    (CCD), 82, 87–8
Compton, Karl T., 43, 56–7,
    144, 147
Communism, 83, 88
Confucianism, 6–7
cosmic-ray physics, 176
cyclotron, 24, 27, 41, 43, 49,
    124, 149

demobilization, 45–8
democracy, 64, 70, 73, 102, 132
    see also laboratory democracy
Democratic Scientists' Association
    (Minka), 60, 81, 87, 100
Democratic Technologists'
    Association, 60
doctorates of science, 9
Dulles, John Foster, 69
Dyer, Henry, 7

Economic and Scientific Section
    (ESS), 11, 61–2
Education, Ministry of, 11, 131, 156
Einstein, Albert, 65, 108
Elementary Particle Theory Group,
    97–102, 125

Federation of Economic
    Organizations (Keidanren),
    70, 151
Feynman, Richard P., 131, 169,
    175–6, 178

F-Project, 38–9
First United Nations International
    Conference on the Peaceful
    Uses of Atomic Energy,
    155, 181
Fox, Gerald W., 49, 149
Fujimoto, Yōichi, 97, 100, 170, 186
Fujioka, Yoshio, 90, 115–16, 121,
    124, 151, 153, 175, 181
Fukuda, Nobuyuki, 169–70, 186
Fushimi, Kōji, 150–1, 165, 181,
    183, 186

Germany, 6
Groves, General Leslie R., 56

Hafstad, Lawrence R., 157
Hani, Gorō, 89
Hayakawa, Satio, 100, 168–95
Heisenberg, Werner, 28, 101, 115,
    120–1
Hevesy, George de, 21, 24, 27

Ichikawa, Ichirō, 157
Ikeda, Hayato, 167, 191
Imperial Academy, 108
    see also Japan Academy
Imperial College of Engineering, 7
Institute for Nuclear Study, 95,
    100, 124–6, 166, 177
Institute of Physical and Chemical
    Research
    see Riken
Institute of Plasma Physics,
    180, 187
Institute of Space and Aeronautical
    Science (ISAS), 175
    see also Institute of Space and
    Astronautical Science
Institute of Space and Astronautical
    Science (ISAS), 193–4
International Conference on
    Theoretical Physics (1953),
    175–6
International Council of Scientific
    Unions (ICSU), 64

International Geophysical Year
    (IGY), 189–90
isotope separation, 38
Itō, Yōji, 37

Japan Academy, 60, 91
Japanese Association for Science
    Liaison (JASL), 58–9
Japan Atomic Energy Commission
    (JAEC), 94, 96, 115–17, 177
Japan Atomic Energy Research
    Institute (JAERI), 116,
    159–60, 162, 168
Japan Atomic Industrial Forum
    (JAIF), 160
Japan Atomic Power Company
    (JAPCO), 158, 162–4, 168
Japan Nuclear Ship Development
    Agency (JNSDA), 162
Japan Society for the Promotion of
    Science (JSPS), 61
Joliot, Frédéric, 93

Kaya, Seiji, 90, 115, 148, 151,
    153, 155
KEK High Energy Physics
    Laboratory, 130, 180
Kelly, Harry, 49, 54, 56, 64, 109, 147
Kigoshi, Kunihiko, 38, 79
Kikuchi, Seishi, 79, 121, 124, 165,
    174, 183–4
Kobayashi, Minoru, 38, 107, 120–1
Kotani, Masao, 122, 175
Kyoto Conference of Scientists, 96,
    139–40
Kyoto University, 78, 107
kyūri, 6

Laboratory Council, 85
laboratory democracy, 77, 86
Lawrence, Ernest O., 23, 27, 40,
    49, 63–4, 143–4, 149, 156
Liberal Democratic Party (LDP),
    114, 129–30
light water reactors (LWRs), 162–4
Lucky Dragon, 154

MacArthur, General Douglas, 49, 69
Manchurian Incident, 22–4
Manhattan Project, 143
Marquat, General William F., 54, 56
Marshak, Robert E., 175–6
Marxism, 73, 82, 89, 101, 132–3, 139
Meiji Restoration (1867–68), 5
Meitner, Lise, 28
Meson Club, 30, 97
meson theory, 108
Minakawa, Osamu, 169, 172–3
Minka
  see Democratic Scientists' Association
mobilization, of science, 31–5
Morrison, Philip
Murray, Thomas E., 151, 155

Nagaoka, Hantarō, 19, 22, 26, 34, 45, 60, 107–8, 120, 146
Nagasaki, 143
Nagoya University, 85–6, 126, 179, 188
Nakamura, Seitarō, 177, 183–4
Nakasone, Yasuhiro, 1, 40, 66–71, 110, 113–14, 118, 128, 152–3, 193, 195, 199
Nambu, Yōichirō, 171, 173–5
National Research Council (NRC), 108
National Science Foundation, 61
National Space Development Agency of Japan (NASDA), 193–5
New Deal, 47
Ni-Project, 38
Nishina, Yoshio, 17–66, 75, 95, 121, 124, 146, 148, 174, 197
Nishida, Kitarō, 133
Nitobe, Inazō, 1, 3
Nobel Prize, 110, 115, 123, 131, 141
nuclear power, 155–65
nuclear fusion, 165–6, 180–8

O'Brien, Brigadier John, 51, 57, 148
Oda, Minoru, 122, 175
Okabe, Kinjirō
Ōkōchi, Masatoshi, 10
Oppenheimer, J. Robert, 56, 63, 110–11
Osaka University, 108

Pais, Abraham, 176
Pash, Colonel Boris T., 56
Peace Treaty, 92
People's Republic of China, 101, 153
Physical Society of Japan, 9, 85, 88, 100, 125
Physico-Mathematical Society of Japan, 9, 97, 107–8
Powell, C.F., 149
Prince Chichibu, 25
Progress of Theoretical Physics, 109, 178
Public Works, Ministry of, 7
Pugwash Conferences on Science and World Affairs, 95–6, 101, 119, 137–9

Rabi, I.I., 21, 52–5
radar, 39–40
Renewal Committee for Science Organization, 59–61
Research Institute for Fundamental Physics (RIFP), 95, 98–9, 124, 183
Riken, 10, 17–18, 20, 25, 55, 62, 76, 97, 147
Rossi, Bruno, 173–4
Russell, Bertrand, 96
Russell-Einstein Manifesto (1955), 95, 139
Russia
  see Soviet Union
Rutherford, Ernest, 20

Sagane, Ryōkichi, 27, 36–7, 79, 143–68, 169, 172

Sakata, Shōichi, 36, 38, 73–7, 81,
    88–91, 93–6, 101–3, 107, 186
samurai, 1, 2, 4
Schwinger, Julian, 29, 131
Science and Technology Agency,
    93, 95–6, 128, 159–61, 193
Science and Technology Council,
    96, 127, 161, 167
Science Council of Japan (JSC), 42,
    60, 70, 85, 90–1, 93–7, 100,
    103, 122, 124, 126, 130,
    152, 161
Science Deliberative Council, 26
Scientific Research Institute,
    55, 124
    see also Riken
Scientific and Technical Division,
    11, 61–2
Sekai bunka (World Culture), 75,
    78–9
Self-Defense Force, 130
Serber, Robert, 143
Shisō no kagaku (The Science of
    Ideas), 82–3
Shōriki, Matsutarō, 95, 116, 118,
    128, 157, 159
Smyth Report, 92
Society for Science for National
    Defense, 26
Solvay Conference, 108
Soviet Union (USSR), 101, 140,
    153, 191
space research, 188–94
Special Committee for Atomic
    Energy, 95–7
Special Committee for Nuclear
    Research (SCNR), 93, 95,
    97–8, 125, 131, 154, 177
Supreme Commander for the Allied
    Powers (SCAP), 73
    see also Allied Occupation of
        Japan; Civil Censorship
        Detachment; Economic and
        Scientific Section;
        MacArthur, General

Douglas; Scientific and
    Technical Division
Suzuki, Tatsusaburō, 36–7, 41

Taketani, Mituo, 36, 42, 73, 76–97,
    99–100, 102–3, 107, 177,
    183–4, 197–8
Takeuchi, Masa, 38–9, 79, 146
Takikawa Incident, 78
Tamaki, Hidehiko, 121
Tamaki, Kajurō, 108, 120
Tanikawa, Yasutaka, 29, 76
Taoism, 132–3
Three Non-Nuclear Principles, 94
Tokyo Bunrika University
    see Tokyo University of Science
        and Literature
Tokyo Mathematical Society, 7
Tokyo Mathematico-Physical
    Society, 7
Tokyo Imperial University
    see University of Tokyo
Tokyo University
    see University of Tokyo
Tokyo University of Science and
    Literature, 122–3
Tomonaga Seminar, 123
Tomonaga, Sin-itirō, 28–31, 79,
    81, 101, 105–6, 111, 119,
    131–3, 135–41, 147, 174,
    195, 199
Truman, President Harry S., 43,
    60–1, 67, 88

Umezawa, Hiroomi, 170, 174
United Nations Educational,
    Scientific, and Cultural
    Organization (UNESCO),
    64, 127
University of Tokyo, 85, 108,
    146–7
U.S. Scientific Intelligence Survey,
    47, 144
U.S. Scientific Advisory Group, 52
U.S. Scientific Mission, 52, 56

Watanabe, Satoshi, 59, 80, 121
Watase, Yuzuru, 122
Willoughby, Maj. Gen.
    Charles A., 68
World Council of Peace, 95

Yamagawa, Kenjirō, 7–8
Yamaguchi, Yoshio, 173
Yasukawa, Daigorō, 155, 158–9
Yoshida, Prime Minister Shigeru,
    68–9, 74, 97

Yukawa, Hideki, 28–9, 33, 38, 41,
    45, 76, 81, 94, 98, 105–12,
    116–18, 132–5, 161, 165,
    173, 177, 183, 185, 191,
    197–8
Yukawa effect, 176
Yukawa Hall, 98, 110, 123
    see also Research Institute for
    Fundamental Physics
    (RIFP)
Yukawa Story, 105

Printed in the United States
By Bookmasters